T0299891

Intensive Longitudinal Analysis of Human Processes

This book focuses on a span of statistical topics relevant to researchers who seek to conduct person-specific analysis of human data. Our purpose is to provide one consolidated resource that includes techniques from disciplines such as engineering, physics, statistics, and quantitative psychology and that outlines their application to data often seen in human research. The book balances mathematical concepts with information needed for using these statistical approaches in applied settings, such as interpretative caveats and issues to consider when selecting an approach.

The statistical topics covered here include foundational material as well as state-of-the-art methods. These analytic approaches can be applied to a range of data types such as psychophysiological, self-report, and passively collected measures such as those obtained from smartphones. We provide examples using varied data sources including functional MRI (fMRI), daily diary, and ecological momentary assessment data.

Features:

- Description of time series, measurement, model building, and network methods for person-specific analysis
- Discussion of the statistical methods in the context of human research
- Empirical and simulated data examples used throughout the book
- R code for analyses and recorded lectures for each chapter available via a link available at www.routledge.com/9781482230598

Across various disciplines of human study, researchers are increasingly seeking to conduct person-specific analysis. This book provides comprehensive information, so no prior knowledge of these methods is required. We aim to reach active researchers who already have some understanding of basic statistical testing. Our book provides a comprehensive resource for those who are just beginning to learn about person-specific analysis as well as those who already conduct such analysis but seek to further deepen their knowledge and learn new tools.

Chapman & Hall/CRC
Statistics in the Social and Behavioral Sciences
Series Editors: Jeff Gill, Steven Heeringa, Wim J. van der Linden, Tom Snijders

Recently Published Titles

Big Data and Social Science: Data Science Methods and Tools for Research and Practice, Second Edition
Ian Foster, Rayid Ghani, Ron S. Jarmin, Frauke Kreuter and Julia Lane

Understanding Elections through Statistics: Polling, Prediction, and Testing
Ole J. Forsberg

Analyzing Spatial Models of Choice and Judgment, Second Edition
David A. Armstrong II, Ryan Bakker, Royce Carroll, Christopher Hare, Keith T. Poole, and Howard Rosenthal

Introduction to R for Social Scientists: A Tidy Programming Approach
Ryan Kennedy and Philip Waggoner

Linear Regression Models: Applications in R
John P. Hoffman

Mixed-Mode Surveys: Design and Analysis
Jan van den Brakel, Bart Buelens, Madelon Cremers, Annemieke Luiten, Vivian Meertens, Barry Schouten, and Rachel Vis-Visschers

Applied Regularization Methods for the Social Sciences
Holmes Finch

An Introduction to the Rasch Model with Examples in R
Rudolf Debelak, Carolin Stobl, and Matthew D. Zeigenfuse

Regression Analysis in R: A Comprehensive View for the Social Sciences
Jocelyn H. Bolin

Intensive Longitudinal Analysis of Human Processes
Kathleen M. Gates, Sy-Miin Chow, and Peter C. M. Molenaar

For more information about this series, please visit: https://www.routledge.com/Chapman--HallCRC-Statistics-in-the-Social-and-Behavioral-Sciences/book-series/CHSTSOBESCI

Intensive Longitudinal Analysis of Human Processes

Kathleen M. Gates
Department of Psychology, University of North Carolina

Sy-Miin Chow
Pennsylvania State University

Peter C. M. Molenaar
Pennsylvania State University

CRC Press is an imprint of the
Taylor & Francis Group, an **informa** business
A CHAPMAN & HALL BOOK

First edition published 2023
by CRC Press
6000 Broken Sound Parkway NW, Suite 300, Boca Raton, FL 33487-2742

and by CRC Press
4 Park Square, Milton Park, Abingdon, Oxon, OX14 4RN

CRC Press is an imprint of Taylor & Francis Group, LLC

Library of Congress Cataloging-in-Publication Data

Names: Gates, Kathleen M., author. | Chow, Sy-Miin, author. | Molenaar, Peter C. M., author.
Title: Intensive longitudinal analysis of human processes / Kathleen Gates, Sy-Miin Chow, Peter C.M. Molenaar.
Description: First edition. | Boca Raton : Chapman & Hall/CRC Press, 2023. | Series: Chapman & Hall/CRC statistics in the social and behavioral sciences | Includes bibliographical references and index.
Identifiers: LCCN 2022022626 (print) | LCCN 2022022627 (ebook) | ISBN 9781482230598 (hardback) | ISBN 9781032354958 (paperback) | ISBN 9780429172649 (ebook)
Subjects: LCSH: Psychometrics. | Psychology--Statistical methods. | Human behavior--Mathematical models.
Classification: LCC BF39 .G38 2023 (print) | LCC BF39 (ebook) | DDC 150.1/5195--dc23/eng/20220914
LC record available at https://lccn.loc.gov/2022022626
LC ebook record available at https://lccn.loc.gov/2022022627

ISBN: 9781482230598 (hbk)
ISBN: 9781032354958 (pbk)
ISBN: 9780429172649 (ebk)

DOI: 10.1201/9780429172649

Typeset in Palatino
by KnowledgeWorks Global Ltd.

Access the support material: www.routledge.com/9781482230598

Contents

Preface

We are pleased to present *Intensive Longitudinal Analysis of Human Processes*. This book focuses on a variety of analytic approaches useful for researchers seeking to conduct person-specific analysis of human data. Person-specific analysis, also referred to as individual-level analysis, allows for statistical inferences for each individual person separately. This individual-level type of analysis is necessary when one might believe that people differ in their dynamic processes, be it emotional, cognitive, behavioral, physiological, or any combination of these. The current thinking is that many of these processes differ in some way across individuals.

Our purpose is to provide one consolidated resource for both understanding and conducting such analysis. In order to provide the necessary information to understand the techniques, we offer details on the math behind the analytic methods. Alongside this technical information, we provide practical guidance on the historical context, application of methods, and interpretation of results.

Foundational information is provided, so no prior knowledge of the specific topics is required or expected. We aim to reach active researchers who already have some understanding of statistical theory and concepts. We provide enough coverage of topics with the aim that those who already carry out such analyses will still learn something from this book, while also being an accessible entry point for those who have never conducted person-specific analyses.

We took great efforts to include statistical methods from a range of disciplines that can be useful in the analysis of individuals. From physics, we describe the notion of ergodicity and how it applies to human studies (see Chapter 2). In Chapters 3, 4, and 5, we draw from traditional time series approaches seen in statistics, sociology, and psychology literature. These are critical for both the measurement of unobservable concepts as well as identifying the lead-lag and instantaneous relations among constructs. Chapter 6 describes a range of model-building approaches that have foundations in economics and computer science. From engineering disciplines, we discuss time-varying parameter estimation (Chapter 7) and control theory (Chapter 8). Chapter 9 focuses on network measures and concepts, many of which originated in the physics literature.

The topics covered here include foundational material as well as state-of-the-art methods. These analytic approaches can be applied to diverse data such as psychophysiological, self-report, and passively collected measures. As such, we provide examples using varied data sources including functional MRI, daily diary, and ecological momentary assessment data. In order to increase the utility of the book, we provide R code so that the reader can practice using these methods. The code as well lectures and other accompanying material for the book can be accessed via a link perpetually available at https://www.routledge.com/9781482230598.

Acknowledgments

The description of methods and demonstration of how to interpret results benefited enormously from the use of empirical data. For this, we thank Aaron Fisher at the University of California, Berkeley, for providing publicly available ecological momentary assessment data. We also have deep gratitude for Kristen Lindquist, a colleague at the University of North Carolina, who generously provided fMRI data for use as an example as well as helped with interpretation. We are also grateful for the contributors of fMRI data to the Autism Brain Imaging Data Exchange (ABIDE), founded by Adriana Di Martino and Stewart Mostofsky.

The published version of the book has benefited greatly from the help of others. Feedback from anonymous reviewers helped to shape the focus of the book.

Katie M. Gates is thankful to the students in her graduate course at the University of North Carolina (UNC), who read early drafts of the book, as well as Adam Miller and Marco Chen, who helped to identify typos and improve clarity in the text. She also is grateful for the efforts of Zack Fisher and Stephanie Lane, who assisted with some of the exemplar code and analysis for use in the book. Additionally, Zack, along with Ken Bollen, provided notes on specific sections pertaining to estimation of SEM. We deeply appreciate this input.

Sy-Miin Chow would like to thank students, colleagues, and collaborators in the Modeling Developmental Systems (MODS) lab, QuantDev, her Dynamical Systems class, and Penn State, for reading earlier drafts of some of the chapters, testing earlier (and ongoing) versions of computer scripts, and serving as audience in some of the recorded lectures included with the book. Part of the book was completed during Chow's sabbatical leave from Penn State, so support from the university and her department, Human Development and Family Studies, is greatly appreciated.

Finally, Peter and Sy-Miin thank John Nesselroade, for his kind mentorship of their work in this field.

Kathleen M. Gates
Sy-Miin Chow
Peter Molenaar

About the Authors

Kathleen (Katie) M. Gates is an Associate Professor of Quantitative Psychology in the Department of Psychology at the University of North Carolina at Chapel Hill. She obtained her PhD in the Department of Human Development and Family Studies (quant focus) at Penn State, a Masters of Forensic Psychology at the City University of New York (John Jay College), and a BS in psychology from Michigan State University. Katie's work is motivated by problems in analyzing individual-level data. She develops algorithms and programs that may aid researchers in better quantifying behavioral, psychophysiological, and emotional processes across time. The end goal is to help researchers identify patterns within individuals so we can provide person-specific prevention, treatment, and intervention protocols as well as better understand the varied basic physiological underpinnings for emotions, cognition, and behaviors.

Sy-Miin Chow is Professor of Human Development and Family Studies at the Pennsylvania State University. She is an elected fellow of the Alexander von Humboldt Foundation in Germany and a winner of the Cattell Award from the Society for Multivariate Experimental Psychology as well as the Early Career Award from the Psychometric Society. Her work focuses on methodologies for handling intensive longitudinal data; methodological issues that arise in studies of change and human dynamics; and models and approaches for representing the dynamics of emotions, child development, and family processes, as well as ways of promoting well-being and risk prevention.

Peter C. M. Molenaar is Distinguished Professor of Human Development and Family Studies at the Pennsylvania State University. He is a recipient of the Pauline Schmitt Russell Distinguished Research Career Award from the College of Health and Human Development at Penn State, the Aston Gottesman Lecture Award from the University of Virginia, the Sells Award for Distinguished Multivariate Research from the Society for Multivariate Experimental Psychology (SMEP), and the Tanaka Award from SMEP in 2017. His work instituted what many characterize as a conceptual and methodological paradigm shift in the analysis of psychological, social, and behavioral processes from an inter-individual to an intra-individual variation perspective.

Notation Used

Δ	change
$\forall$	all inclusive
Λ	matrix of factor loadings
Ψ	covariance matrix of latent variables (or latent variable errors)
Ξ	covariance matrix of measurement errors
$\mathbf{B}$	beta matrix or backshift operator (distinction explicit in text)
Σ	population covariance matrix
$\mathbf{C}$	sample covariance matrix
$\mathbf{y}$	observed variable
$\mathbf{\Phi}$	auto- and cross-regression parameters
Θ	moving average parameters
p	indicates the total number of observed variables
q	indicates the total number of latent variables
i	individual (person) indicator
N	total number of individuals
t	time indicator
T	total number of observations across time
m	lag order for autoregressive model components
n	lag order for moving average model components

Scalars will be indicated with *italicized* letters,
vectors will be denoted by **bold face lowercase** letters,
and matrices by **BOLD FACE UPPERCASE Greek or Roman** letters.

List of Abbreviations

ACF:	Autocorrelation function
ADID:	Affective Dynamics and Individual Differences
AIC:	Akaike Information Criterion
AR:	Autoregressive
BIC:	Bayesian Information Criterion
CDEKF:	Continuous-Discrete Time Extended Kalman Filter
CTT:	Classical Test Theory
DST:	Developmental Systems Theory
EEG:	Electroencephalogram
EKF:	Extended Kalman filter
EKS:	Extended Kalman smoother
EMA:	Ecological momentary assessment
ESM:	experience sampling methods
FIML:	Full-information maximum likelihood
fMRI:	Functional Magnetic Resonance Imaging
GARCH:	Generalized autoregressive conditional heteroskedastic
GIMME:	group iterative multiple model estimation
gVAR:	graphical VAR
IAV:	intra-individual variability
IEV:	inter-individual variability
IRW:	Integrated random walk
LQC:	Linear Quadratic Controller
MIIV-2SLS:	model implied instrumental variables estimated with two stage least squares
MLE:	Maximum likelihood estimate
MRI:	Magnetic Resonance Imaging
MRSRM:	Multivariate Replicated Single-Subject Repeated Measurement
PED:	Prediction Error Decomposition
RHC:	Receding Horizon Control
RMSD:	Root Mean Squared Deviation
RW:	Random walk
SVAR:	structural VAR
TVP:	Time-varying parameters
uSEM:	uni!ed Structural Equation Modeling
VAR:	vector autoregression

1

Introduction

1.1 First Encounter with Intra-Individual Variation

Variation is the lifeblood of inductive science, in particular statistics. It is therefore appropriate to begin this book with a standard dictionary definition of variation:

> *The degree to which something differs, for example, from a former state or value, from others of the same type, or from a standard.*

The initial part of the definition, "The degree to which something differs ...", implies making a comparison. The remaining part of the definition then specifies three types of comparison: (1) between different temporal states of a given something, (2) between different replicates of the same type of something, or (3) between the given something and a standard. As an example, type-1 variation can be seen when looking at moods across time within people and type-2 variation can be seen when looking at moods across people. Type-3 variation can be used within each of the prior two. For instance, one can look at variation within a person from their starting point, or one can look at variation between people in reference to a specific score. These three types of comparison each reveal variation, but as will be shown in what follows, there are fundamental differences between variation revealed by the first and second types of comparisons (we will not further discuss the third type of comparison as it can be interpreted as special cases of the first types). That is, considering variation obtained by comparing across different temporal states of a given something differs in fundamental respects from variation obtained by comparing across different replicates of the same type of something.

Intra-individual variation (IAV) in its most basic form is a type-1 variation, resulting from a comparison between different temporal states of a given something. For instance, IAV describes changing moods across time for an individual, or change in brain activation across time for a given individual. In principle, the restriction to temporal states in the dictionary definition given above can be extended to spatio-temporal states (for instance of an

DOI: 10.1201/9780429172649-1

electro-dynamic brain field) or replaced by spatial states (for instance, the number of glial cells across brain regions). But we will primarily restrict attention to temporal variation because that is the most common form encountered in the social and bio-behavioral sciences using intensive longitudinal data. For the purposes of this book, IAV is defined as the study of variability across time for individuals. Psychologists often aim to describe individuals in their research but less frequently collect and model their data in a way that can provide inferences about individuals; intensive longitudinal analysis of IAV meets this need.

Inter-individual variation (IEV), which is examined in cross-sectional analysis, in its most basic form is a type-2 variation, resulting from a comparison between different replicates of the same type of something. Carrying on with the previous examples, we can examine variability across individuals in their mood, or how activation in a brain region varies across individuals. Again, the dictionary definition can be extended in that different replicates can be compared at a number of consecutive measurement occasions. As will be explained in what follows, this so-called longitudinal variant of IEV has given rise to much misunderstanding about its relation to IAV. The reason is that consideration of IEV across a sequence of measurement occasions is suggestive of combining type-1 and type-2 variations. Unfortunately, this impression is incorrect. Even in the longitudinal case, IEV is fundamentally the study of how individuals vary in comparison to each other (we present a toy example in the online supplement). In this chapter, we separate the notions of IAV and IEV, describe its roots in classical test theory, and provide numerous examples of IAV across a number of disciplines within psychology. We then provide an overview of the topics to come in the book.

1.1.1 Cattell's Data Box

A convenient heuristic device to further highlight the differences between IAV and IEV is Cattell's (1946) data box (see Figure 1.1). The axes correspond to persons, variables, and occasions. In terms of this data box, IAV in its simplest form is defined by the following three steps:

- IAV1. Select a given individual. Inferences are focused on that one individual, meaning that no generalization across individuals is considered.
- IAV2. Select a set of variables. Inferences are focused on that one set of variables.
- IAV3. Select a set of multiple measurement occasions. Generalization across the occasion axis is considered.

Following these three steps yields the cross-section called INTRA-individual Data depicted in the right-upper corner of Figure 1.1. According to

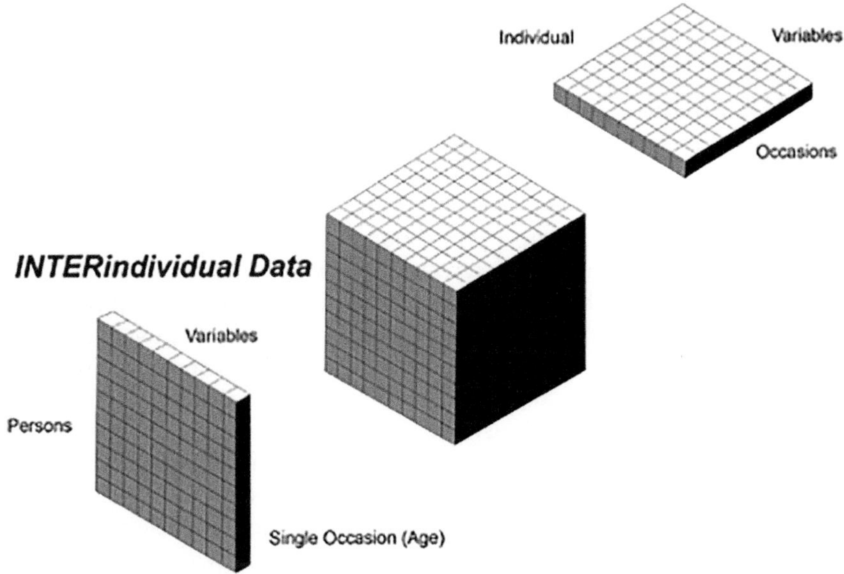

FIGURE 1.1
Cattel data box.

step IAV3 above, the set of measurement occasions is "random". The standard way to interpret this is as follows. The first and last measurement occasions determine an interval of time. This interval is considered to be randomly drawn from the set of all occasions, where the latter set is spanning time from negative infinity to infinity. Obviously, this is an idealization, but it makes clear that it is the complete interval that is randomly sampled from the population of all time intervals for a given person. It is noted that other interpretations are possible, for instance, taking time to start at zero and running until infinity. But for the moment this is not further elaborated.

IEV is defined in terms of the data box by the following three steps:

- IEV1. Select a set of multiple subjects. The set of subjects is considered to be random, meaning that generalization to the population of subjects is considered.
- IEV2. Select a set of variables. Inferences are focused on that one set of variables.
- IEV3. Select one measurement occasion. This occasion is regarded as being "fixed", meaning that no generalization across the occasion axis is considered.

Following these three steps yields the cross-section called INTER-individual Data in the left-lower corner of Figure 1.1.

The given definitions of IAV and IEV focus on the formal properties of these two types of variation. It is possible to further characterize IAV and IEV, for instance, by distinguishing sub-types within each main type. An interesting example of such further characterization is presented in Li et al. (2004), where four sub-types of IAV are introduced. Each sub-type is a function of the time scale at which the purported variation unfolds (fast or slow) and depends on whether this variation is associated with cumulative or reversible processes. However, most of the contents of this book pertain to the general concept of IAV and therefore further characterizations of IAV can be neglected.

In the next chapter, the comparison between IAV and IEV will be continued in considerable depth. But in the remainder of this chapter, the focus is on IAV. It is evident that analysis of IEV has been much more prominent in quantitative psychology and psychometrics than analysis of IAV. The aim of much data analysis in psychology has been on generalizing results obtained in analyses of IEV to some populations or sub-populations of individuals, and therefore analysis of IEV can still be referred to as the standard approach to data analysis in psychology. But there exist particular sub-disciplines in psychology and related fields of research in which analysis of IAV traditionally has fulfilled a prominent role. It is to some of these fields that have traditionally utilized analysis of IAV that we now turn.

1.1.2 IAV in Psychology and Related Sciences

The first sub-discipline of psychology where IAV fulfills an important role is *classical test theory* (CTT). This may come as a surprise to those who know about and/or apply CTT, because it appears to be firmly based on the analysis of IEV. But it will be shown that IAV plays an important theoretical role in the foundations of CTT. What follows is mainly based on Molenaar (2008).

The basic concept in CTT is the concept of true score: each observed score is conceived of as a linear combination of a true score and an error score. In their authoritative book on CTT, Lord and Novick (1968) define the concept of true score as follows. They consider a fixed person P, i.e., P is not randomly drawn from some population but is the given person for which the true score is to be defined. The true score of P is defined as the expected value (i.e., the mean) of the propensity distribution of P's observed scores. The propensity distribution is characterized as a

> *... distribution function defined over repeated statistically independent measurements on the same person.*
>
> (Lord & Novick, 1968, p. 30)

The concept of error score then follows straightforwardly: The error score is the difference between the observed score and the true score.

Several aspects of this definition of true score are noteworthy. The definition is based on IAV characterizing a fixed person P. Repeated administration of the same test to P yields a time series of scores of P, the (presumably constant) mean level of which is defined to be P's true score. Hence this definition of true score does not involve any comparison with other persons and therefore is not dependent on IEV. Lord and Novick (1968) require that the repeated measurements be independent. This implies that the time series of P's scores should lack any sequential (over-time) dependencies, meaning that a score at a given time point is not predicted by scores at prior time points. For instance, weight may fluctuate from day to day based on food intake, hydration level, and other factors and may not be predicted by the prior day's weight but rather fluctuate around the person's average weight. However, if we looked within a day, we might see that the prior time point's weight is predictive of the next time point. The Appendix provides an introduction into the consideration of time series approaches using the Lord and Novick conceptualization of CTT as a guide.

Lord and Novick (1968, p. 30) do not further elaborate their original definition of true score in the context of IAV because

> *... it is not possible in psychology to obtain more than a few independent observations.*

Instead of considering an arbitrary large number of replicated measurements of a single fixed person P, Lord and Novick (1968, p. 32) shift attention to an alternative scheme in which an arbitrary large number of persons is measured at a single fixed time:

> *Primarily, test theory treats individual differences or, equivalently, the distribution of measurements over people.*

In other words, they suggested that the propensity distributions associated with individual true scores can be illuminated by using an individual differences approach.

Before focusing in the remainder of their book solely on the latter definition of true score based on IEV, Lord and Novick (1968, p. 32) make the following interesting comment about their fundamental definition of true score based on IAV:

> *The true and error scores defined above* [sc. based on intra-individual variation] *are not those primarily considered in test theory ... They are, however, those that would be of interest to a theory that deals with individuals, rather than with groups (counseling rather than selection).*

This is a remarkable, though somewhat oblique statement. What is clear is that Lord and Novick consider a test theory based on their initial concept of true score, defined as the mean of the IAV characterizing a fixed person P, to be "... of interest to a theory that deals with individuals ...". That is, they consider such a test theory based on IAV to be important in the context of individual assessment. But what is not clear is whether they also consider the alternative concept of true score based on IEV (individual differences) to be not of interest to a theory that deals with individuals. That is, do they imply that CTT as we know it is only appropriate for the assessment of the selection of individuals from groups and not for understanding individuals?

To summarize, Lord and Novick (1968) initially defined the fundamental concept of true score as the expected value of the propensity distribution of the observed scores of a fixed individual person P. This definition of true score based on IAV then is transposed to an IEV context focused on individual differences, i.e., CTT as we know it. In the next chapter, we will return to CTT and further discuss this remarkable shift from an initial IAV perspective to an IEV perspective in the theoretical underpinnings of CTT.

1.1.3 In What Areas Have the Studies of IAV Been Useful?

Perhaps the oldest tradition in psychology associated with a focus on IAV is the *idiographic* approach. Allport (1937) popularized the idiographic-nomothetic distinction in personality psychology, defining the aim of idiographic approaches to identify patterns of behavior, thought, and emotion within an individual over time and of contexts. In contrast, Allport (1937) defined the nomothetic approach in terms of a focus on IEV. The original idiographic-nomothetic distinction is due to the German neo-Kantian philosopher Windelband (1901) and did not refer to different types of variation and methods to analyze these, but instead, to different types of knowledge. According to Windelband, idiographic knowledge aims at describing and explaining particular phenomena, whereas nomothetic knowledge is about generalities that are common to a class of particulars. For instance, the Schrödinger wave equation constitutes an example of nomothetic knowledge in that it applies to all non-relativistic quanta. Robinson (2011) emphasized the importance to not confound Windelband's original definition in terms of kinds of knowledge with Allport's later definition in terms of kinds of variation. That is, Robinson argues that nomothetic knowledge in the sense intended by Windelband has no relation with IEV, but rather, refers to lawful knowledge that equally applies to all elements of a class of otherwise unique phenomena. Specifically, IEV often attempts to differentiate individuals based on covariation in constructs across individuals. Windelband's conception of nomothetic instead would be research that aims at describing a process that applies to all individuals.

We agree with Robinson that the difference between the definitions of the idiographic-nomothetic dichotomy by Windelband and Allport is important. But within social science, and psychology in particular, it is Allport's definition in terms of IAV versus IEV that has been the most influential. Hence this is the definition that will be adhered to in this book. Our take in this book is that the idiographic approach can be understood as a special case of IAV analysis.[1] The contrasting concept of a nomothetic approach will not be restricted to IEV, but will be used in the more general sense stipulated by Windelband (which includes analysis of IEV as a special case).

This brings us to some distinctions in terminology. In general, the term "person-specific" has been used in psychology interchangeably with "idiographic". That is, both are used to indicate that the focus of analysis is at the level of the individual. Between-person, IEV, or cross-sectional analysis is not used or, if it is, is not focal to the research questions at hand. An additional term, "person-oriented", refers to a specific approach which overlaps greatly with person-specific approaches yet carries additional properties (Bergman & Magnusson, 1997). Idiographic can be considered a general term, under which person-oriented is a specific type of an idiographic approach. A key distinction is that the person-oriented approach presumes that dynamic processes are partly specific to individuals with some characteristics being common across all individuals (as well as across all time points for a given individual). Using terminology such as idiographic and person-specific does not carry these properties as part of their assumptions and is agnostic to whether dynamic processes share some common elements across individuals.

Increasingly, researchers are collecting real-time self-reports or indices of behavior, cognition, or emotions. Equivalently referred to as "ecological momentary assessments" (EMA; Stone & Shiffman, 1994) or "experience sampling methods" (ESM; Csikszentmihalyi & Larson, 1987), these intensive assessment techniques provide data by which a research participant can be assessed close in time to experience. Usually, numerous time points are collected across a sample size that is comparable to those seen in IEV analysis. As such these data collection approaches are considered to enable modern variants of the idiographic method (Conner et al., 2009; Trull & Ebner-Priemer, 2009).

Behaviorism and its current version behavior analysis also focused on IAV, in particular on individual learning curves obtained under various experimental conditions (e.g., Bouton, 2007; Pierce & Cheney, 2013). It should be acknowledged, however, that dedicated statistical analysis of learning curves in this field has not been prominent. To determine whether an experimental manipulation had an impact, eyeballing the graphs of learning curves often is considered sufficient. However, such visual analysis may be problematic for a number of reasons, for instance, due to the presence of sequential correlation in the data (cf. Ballard, Hinton, & Sejnowski, 1983). Relying solely upon the visual analysis of IAV will not be considered further in this book (although visual depictions of intra-individual data are provided, and examination of those prior to analysis is encouraged).

Cognitive neuroscience or psychophysiology is another discipline in which IAV plays important roles. Perhaps because these types of data often contain substantial numbers of time points for each individual, psychophysiological

---

[1] However, IAV needs not be confined to single-subject analysis per se. We will consider this in later chapters.

data has historically been modeled at the individual level. Even from its inception, analysis of electroencephalogram (EEG) data has inherently been individual-level. In EEG, the temporal processes of the brain are modeled within each individual to obtain values regarding the cyclicity of the brain functioning (e.g., the power of gamma and alpha waves) as well as event-related potentials for individuals across a task design. Similarly, the power of certain frequencies found in heart rate data is obtained for individuals separately. In this way, only the individuals' temporal dynamics are considered with no influence from other individuals. In functional MRI (fMRI), researchers are increasingly using time series analysis to model brain functioning at the individual level after attempts in arriving at aggregate models have failed to describe the participants in the study (see Finn et al., 2015; Miller et al., 2002).

Some prominent approaches in *developmental psychology* focus on IAV. Wohlwill's (1973) landmark monograph on the study of behavioral development puts forward the concept of developmental function as fundamental for developmental psychological research. According to Wohlwill

> *... developmental functions refer to the pattern of change displayed by the individual, and can only be determined through successive observations on the same individual.*
>
> (Wohlwill, 1973, p. 33)

A related point of view is put forward by, e.g., Gottlieb (1992), who proposes a developmental systems model involving bi-directional genetic, neural, and behavioral influences to explain individual development. From this developmental system IAV perspective Gottlieb has formulated a fundamental critique of developmental behavior genetic approaches based on analysis of IEV (Gottlieb, 1995, 2003).

The theoretical perspectives of Wohlwill, Gottlieb, and others can be conceived of as variants of the overarching *Developmental Systems Theory* (Ford & Lerner, 1992; Johnston & Lickliter, 2009; Lerner, 2002; Overton, 2014). Developmental Systems Theory conceptualizes development as the result of multiple co-acting influences which are context sensitive and contingent. This implies that development is inherently subject-specific and stochastic. A second important feature of Developmental Systems Theory is that development is understood to be a process in which nonlinear epigenetic influences play central roles (cf. Bateson, 1972; Lickliter & Honeycutt, 2009). Such nonlinear epigenetic influences create substantial subject-specific IAV that reinforces the subject-specific effects due to contingent contextual influences (Ho, 2013; Saunders, 2013). A third important feature of Developmental Systems Theory is its focus on the potential for IAV evolving at multiple time scales and at multiple levels (e.g., Newell et al., 2010; Smith & Thelen, 2003). For an overview of advanced topics in Developmental Systems Theory, the reader is referred to Molenaar et al. (2014).

A focus on IAV also exists in different fields of research in *experimental psychology*. This is especially the case in those parts of experimental psychology dealing with mental chronometry (Jensen, 2006) and human decision-making (Luce, 1999). The key measure in mental chronometry is reaction time, i.e., the elapsed time between the presentation of a sensory stimulus and a behavioral response. The standard experimental design used in mental chronometry is to repeatedly deliver the stimulus to the same individual subject, yielding a sequence of reaction times for this subject. Often this is replicated for a small number of experimental subjects. The distribution and sequential dependencies of this sequence then are determined for each subject and used to infer the mental organization underlying the generation of the responses (Luce, 1999; Townsend & Ashby, 1984). Research on human decision-making can be conceived of as a generalization of choice reaction time experiments in which the focus is on the particular alternative that is chosen from among a range of alternatives. The data obtained in experiments in human decision-making often show extreme degrees of variability across participants. To accommodate these large individual differences, Luce recommends the following IAV approach:

> *Whenever we look, we find substantial individual differences, so aggregation must be done with a good deal of restraint and care. Thus, this monograph, which attempts to deal with scientific questions, focuses attention whenever possible on individual axioms tested on individual people.*
>
> (Luce et al., 1999, p. 29)

In several other scientific fields, there is a focus on IAV, or better, intra-system variation. *Econometrics* is a prime example where this is the case (e.g., Martin, Hurn, & Harris, 2012). In particular, macro-economic modeling is based on time series analysis, often using special time series models such as generalized autoregressive conditional heteroskedastic models (GARCH; Bollerslev, 1986). The range of dynamic modeling approaches in econometrics is quite large, encompassing also dynamic factor models and state-space models. We will pay ample attention to these models in this book. The importance of analysis of intra-system variation in *physics* and *engineering*, usually called signal analysis, is also huge. A classic early monograph is by Papoulis (1977). Particular fields of physical research that use modeling approaches with some affinity to the models discussed in this book are antenna theory (e.g., Drabowitch et al., 2010) and geophysical signal analysis (starting with the classic monograph by Robinson & Silvia, 1981).

In closing this overview of traditional fields of application of analysis of IAV, we should mention the emerging field of *personalized medicine* in which medical decisions, practices, and medication type/dose are tailored to the individual patient. Especially relevant are so-called dynamic treatments of chronic diseases such as hypertension, diabetes, mental illness, obesity, addiction, etc. Clinicians treat these chronic patients by individualizing the treatment type, dosage, and timing according to ongoing assessments of patient response, adherence, side effects, etc. Instead of considering a single treatment, clinicians sequentially make decisions about what to do next to improve each patient's outcome (dynamic treatment; cf. Chakraborty & Moodie, 2013). This requires repeated diagnostic testing of each individual patient in order to adaptively optimize medical treatment in real time. Schwartz and Collins (2007) suggest the use of nano-sensors, molecularly imprinted polymers, and portable electronic devices to continuously monitor environmental stressors in studying gene-environment interactions associated with disease. The spectacular progress in creating such new sensor methods yielding an abundance of IAV data streams will further push forward the application of personalized medicine, as will be shown in our illustration involving the application of dynamic treatment to patients with type-1 diabetes.

1.2 Statistical Analysis of IAV: An Overview of the Structure of This Book

From a strictly statistical perspective, the analysis of IAV amounts to time series analysis. Analysis of IAV in the physical and engineering sciences usually is called signal analysis, whereas in the psychological sciences, the use of "intensive longitudinal data analysis" has emerged. Henceforth the terms "intensive longitudinal analysis" and "time series analysis" will be used interchangeably to describe IAV. There exists an enormous and continuously growing literature on time series analysis. In this section, we only can provide a few pointers to this literature for the advance learner seeking more technical documentation. The main theme of this section is to give a general (and necessarily simplifying) characterization of the existing literature on time series analysis and distinguish the contents of the present book with respect to that literature.

Time series analysis in its present form is of relatively recent origin. Perhaps Yule (1926) can be regarded as the founding paper on modern time series analysis. The reader is referred to Klein (1997) for an excellent overview of the early history of time series analysis, culminating in Wold's (1938) classic monograph on stationary time series. Consecutively, important work in the foundations of the mathematical-statistical theory for stochastic processes was carried out by Kolmogorov (1933) and Wiener (1930). Brillinger (1975, section 2.11) gives a concise but insightful discussion of this work, relating the probability theoretical approach of Kolmogorov to the functional approach of Wiener. In this book, we will follow the probabilistic approach of Kolmogorov (1933). Additional landmark contributions to the fundamental theory of stochastic differential equations were made by Lévy (1937), Itô (1944), and Stratonovich (1966).

Modern time series analysis really took off with the appearance of four landmark textbooks: Jenkins and Watts (1968), Box and Jenkins (1970),

Hannan (1970), and Brillinger (1975). Only Box and Jenkins (1970) presents time series analysis mainly in the time domain, whereas the remaining three textbooks focus on time series analysis in the frequency domain. Time series in the time domain is the focus of this book. Another noteworthy difference is that Box and Jenkins (1970) focuses on univariate time series, either on themselves or as a function of measured univariate input series, whereas the other three textbooks discuss multivariate time series of arbitrary (finite) dimension. Last but not least, control theory only is explicitly addressed in Box and Jenkins (1970). None of these textbooks discusses Bayesian analysis, which is especially noteworthy for the Box and Jenkins' (1970) textbook in that its first author also published a well-known monograph on Bayesian inference (Box & Tiao, 1973). All four textbooks only consider time series observed at discrete equidistant time points.

We will not attempt to summarize here the manifold developments in time series analysis that took place after this important initial work. In the remainder of this book, many references to these further developments will be given within the context of the ongoing presentation. Presently, we will summarize in what respects the contents of the present book differ from the existing literature. It should be acknowledged that these differences only are partial because there also is considerable overlap with modern textbooks on time series analysis. But the unique combination of special emphases in this book sets it apart from the existing literature on time series analysis.

1.2.1 Focus on Dynamic Factor Models

Almost every important psychological construct can be indirectly assessed using multiple indicators. Therefore, the emphasis in this book is on multivariate time series, where each univariate component series is associated with repeated measurements of a distinct indicator or a set of indicators. The main modeling tools are dynamic factor models, i.e., suitable dynamic generalizations of factor analytic models of IEV. Factor models are the most important statistical models in psychology. The reader is referred to Mulaik (1987) for an interesting historical discussion of the foundations of factor analysis. Modern factor modeling including special variants like latent growth curve and other multilevel models, general longitudinal factor models and multi-group factor models are central to psychological data analysis of IEV (e.g., Hoyle, 2012). It therefore would seem appropriate to focus on dynamic factor models in psychological data analysis of IAV. It is expected that this will ease the accessibility of multivariate time series modeling to quantitative psychologists and psychometricians who are well acquainted with standard factor modeling of IEV. It also will allow for direct comparisons between factor analysis of IEV and IAV using the same psychological measures (indicators).

The most important reason to focus on dynamic factor models, however, is that a wide range of multivariate time series models can be conceived of as special variants of dynamic factor models. For instance, the commonly used *state-space models* will be shown to be a specific instance of the general dynamic factor model. Additionally, factor analysis conducted across individuals can be considered a type of dynamic factor model (as introduced in Chapter 3). In fact, all linear time series models can be reformulated as dynamic factor models (e.g., Durbin & Koopman, 2012; Hannan & Deistler, 1988). This also is the case for a wide range of nonlinear time series models, as will be explained at some length in Chapter 7.

1.2.2 Focus on Replicated Multivariate Time Series

Time series analysis in its purest form concerns the statistical analysis of repeated measurements obtained with a single subject or system. An example of the first few rows of a time series data set is depicted in Figure 1.2.

These data are depicted in Figure 1.3. Keep in mind that these are the variables for one person across time.

This focus on single-subject time series characterizes most of the literature on time series analysis, including the classic textbooks referred to above. Single-subject multivariate time series analysis also occurs in psychology, a famous early example being the analysis of a schizophrenic patient's IAV on some cognitive variables (word association and perceptual speed) and a biochemical variable (creatine) across 245 successive days (Holtzman, 1963).

Despite the opportunity to analyze data at the individual level, analysis of IAV in the psychology and bio-behavioral science often is based on multi-subject or multi-system time series. The main reason for the relative popularity of this so-called replicated time series design is the traditional focus in social scientific research on generalizability across some populations of subjects. Although the replications (i.e., individuals) are considered to be fixed, the focus often is on commonalities of important features across the replications. If these are indeed found, then they are interpreted as possible nomothetic relationships that hold for a population at large. Nesselroade and Ford (1985) are early proponents of this approach, calling it the Multivariate Replicated Single-Subject Repeated Measurement (MRSRM) design. In this book, we provide methods for individual-level analysis in addition to approaches for arriving at group-level inferences.

	Lazy	Dynamic	Selfish	Unimaginitive	Irritable
Day 1	15	53	38	15	10
Day 2	12	49	37	6	27
Day 3	12	43	25	14	11
Day 4	27	38	20	9	20
...	.	.	.	.	.
Day T (last day)	24	37	35	8	14

FIGURE 1.2
Example of time series data for daily self-reports. Simulation of data gathered across a total of *T* days. Dots indicate the passage of time/omitted observations for demonstrative purposes.

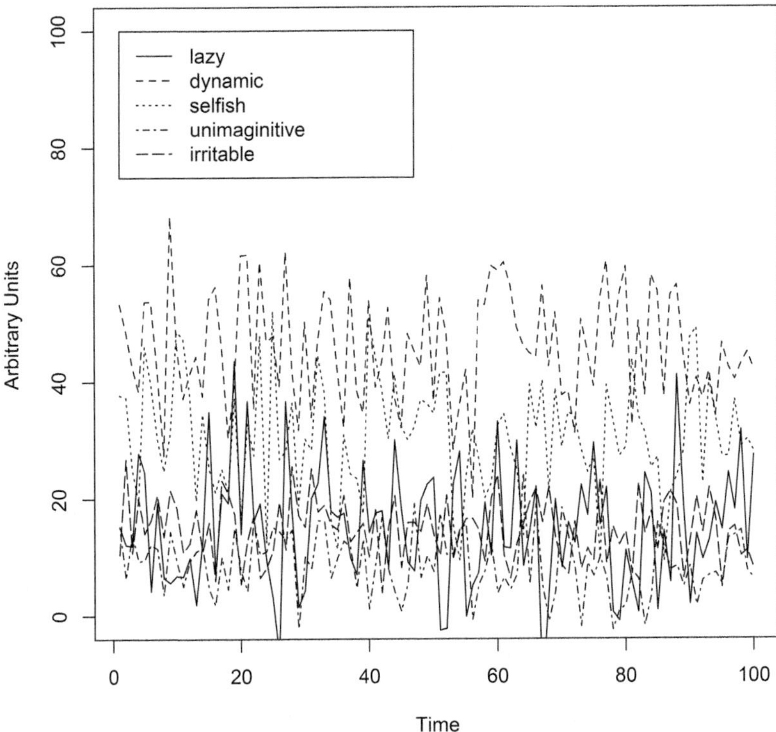

FIGURE 1.3
Example of time series analysis plotted. The x-axis indices time and the y-axis the value of the self-reported variable at each time point.

1.2.3 Focus on User-Friendly Model Selection and Estimation Approaches

The approaches considered in this book can currently all be implemented with R packages (R Core Team, 2014). The usage and output will be described and applied in illustrations presented in this book. The corresponding code is made available in the online supplement via a link found at https://www.routledge.com/9781482230598, including documentation and some additional examples. Following along with the online supplement requires that users spend some extra effort to get acquainted with R software. However, we pay attention to user-friendly approaches, and much of the analyses described in the foundational Chapters 3–5 can be carried out with standard structural equation modeling software. If no prior information is available about an observed p-variate time series, several data-driven procedures specific to R are considered (see Chapter 6). Applications of the pertinent exploratory procedures to real data are presented as illustrative examples throughout the book.

This book serves to provide illustrative examples and step-by-step guidelines on conducting IAV analysis, with special emphasis on variants of dynamic factor models with multiple-subject data. Some of the models and analytic steps in Chapters 2–5 provide a foundation for readers new to IAV analysis: Ergodicity, P-Technique, Vector Autoregression, and Dynamic Factor Modeling. It is highly recommended that these first five chapters are read in order, as one chapter builds from the previous. Several innovative extensions and variants of the dynamic factor model are presented in Chapters 6–9, which we highlight here briefly. These chapters may be read in any order once the reader feels comfortable with concepts in the first five chapters. Each chapter in this book, excluding the present one, contains simulated data examples followed by empirical examples.

1.2.4 Special Topic: Methods for Dealing with Heterogeneous Replications

Analysis of replicated time series almost always shows that the replications are heterogeneous, implying that the data obtained with different subjects obey different, subject-specific, dynamic models. It will be shown that this kind of heterogeneity is difficult to detect in the analysis of IEV. Nonetheless, its predominance in human populations is suggested by theoretical results obtained in mathematical biology and, to reiterate, is corroborated by the results of many analyses of empirical replicated time series data. The apparent ubiquity of such heterogeneity jeopardizes the main goal of using replicated time series designs, namely, to uncover commonalities across replications. Thus, it can be difficult to uncover patterns that are common across replications due to the high degree of individual-level variability in models.

In Chapter 6, several alternative methods to deal with heterogeneous replications will be considered, including a state-of-the-art method that has been proven to be extremely successful in validation studies with simulated data as well as in empirical applications. It is also shown that averaging across even mildly heterogeneous replications can give quite invalid results, such as estimated phantom (or spurious) relations that do not describe the dynamics of any subject in the sample. Further extensions will be introduced that enable both the factor loadings and dynamic relations among latent constructs to vary across individuals.

1.2.5 Special Topic: Non-Stationary Dynamic Factor Models

As will be explained in Chapters 2 and 4, basic time series analysis starts with the assumption that the observed series is stationary. This implies that the time-dependent statistics characterizing the observations obey strict invariances across time. For instance, the time-dependent means have to be constant across all times, which is called "stationarity of the mean" or "first-order stationarity" (because the mean is the first-order moment of the probability distribution characterizing the data). In addition, the variances have to be constant across all times, which is called "stationarity of the variances" or "second-order stationarity" (note that second-order stationarity also refers to stationarity of the sequential covariances; see further details in Chapters 2 and 4), and so on.

The stationarity assumptions in the analysis of IAV fulfill the same role as the homogeneity assumption in the analysis of IEV in that they enable the construction of appropriate estimators. This will become apparent in Chapter 2. Yet they are often violated to various degrees in empirical data, particularly in social scientific and bio-behavioral assessments. The possible sources of non-stationarity are manifold. For instance, attentional, habituation, learning, and developmental processes all are potential sources of time-varying effects that evolve at different time scales. This can result in non-stationarity of the mean function, or the variance function, or a subset of parameters describing the sequential dependencies of the data. Consequently, we will give ample attention to various statistical methods dealing with non-stationarity. These methods range from elementary ways to detect non-stationarity (Chapter 4) up to advanced statistical techniques to blindly (i.e., without any prior knowledge available) fit dynamic factor models to possibly non-stationary multivariate time series in Chapter 7.

1.2.6 Special Topic: Control Theory

Analysis of intra-system variation is often coupled with *optimal control* in engineering applications. If the data gathered about a dynamic system includes time-varying external input that can, at least to some degree, be manipulated by a controller then this input can be used to steer the system in such a way that some performance criterion is optimized. Control theory has become a very sophisticated and wide-ranging field of research, including different types of control (e.g., feedback, feedforward, feedback-feedforward, fuzzy, chaotic) based on various mathematical design principles (linear-quadratic-Gaussian, H-infinity, adaptive). Good modern references are Goodwin and Sin (1984) and, in particular, Kwon and Han (2005), who present detailed derivations. Control theory also has become an important approach in econometrics. As to that, a key reference is Hansen and Sargent (2008), who extend earlier work by Whittle (1990). In this book, we will further discuss and illustrate the importance of control theory in the analysis of IAV, including an ample explanation of the basic principles involved, in Chapter 8.

1.2.7 Special Topic: Intersection of Network Science and IAV

Network science provides a wide range of measures and techniques, some of which can be applied to estimates obtained from time series analyses. The measures quantify aspects of the output matrix to give insight into concepts such as how inter-related the objects in the system are (e.g., variables) and which objects tend to be most central to the overall system. In terms of algorithms, *community detection* is a class of methods for subsetting objects (such as variables or people) into subgroups (see Fortunato, 2010, for a comprehensive review of foundational approaches). We describe popular measures and methods that have been evaluated with simulation studies, applicable to time series analyses, and are widely used in the human sciences. Throughout the chapter, we draw comparisons to approaches likely familiar to the reader.

1.3 Description of Exemplar Data Sets

Clearly, IAV applications can be quite varied—from studying moods and behaviors across time *in vivo* to the psychophysiological relations occurring within a short time frame. To capture this diversity in intensive longitudinal data collected on humans, several empirical data sets are used in the examples in this book. Some also are available as online supplements along with exemplar code. These include two sets of daily assessment data, one set of ecological momentary assessment data and one set of fMRI time series obtained in a resting state.

1.3.1 Big Five Personality Daily Data

Borkenau and Ostendorf (1998) collected daily data on 22 German undergraduate students (19 females) who were predominantly aged 19 or in their early 20s (with the exception of three females that were in their 30s or 40s). The data were collected explicitly to identify if the longitudinal relations of these traits evidenced similarities across individuals or if the individuals differed in their dynamic processes. Every evening for 90 days, participants indicated how well 30 adjectives (6 for each personality trait) described their behavior that day on a scale of *0 = not at* all to *6 = extremely.* The personality traits were as follows: Neuroticism, Extraversion, Agreeableness, Conscientiousness, and Intellect. The adjectives, which were in German, had previously been tested and identified as appropriate markers for the specific traits (Ostendorf, 1990). Furthermore, the adjectives were ones that are considered to be descriptive of either a state or a trait, thus facilitating comparison of between- and within- individual results. The data has been explored elsewhere using IAV analysis (e.g., Hamaker, Dolan, & Molenaar, 2005; Molenaar & Campbell, 2009), and we describe these analyses and results in Chapters 3 (P-technique) and 4 (Vector Autoregression).

1.3.2 Fisher Data

Fisher and colleagues (Fisher et al., 2017) made datasets of time series for 40 individuals publicly available. The individuals were assessed four times a day for a minimum of 30 days, with each individual in the shared data having at least 110 time points. Participants were asked to rate symptomatology using a 0–100 visual analog slider on 21 items drawn from the diagnostic criteria for Generalized Anxiety and Major Depressive Disorders. We focus on this data in Chapters 6 (Model Building) and 9 (Networks).

1.3.3 The ADID Study

The Affective Dynamics and Individual Differences (ADID; Emotions and Dynamic Systems Laboratory, 2010) study was composed of 217 participants whose ages ranged between 18 and 86 years old enrolled in a laboratory study of emotion regulation, followed by an EMA study during which the participants rated their momentary feelings 5 times daily over a month. This data set has been published previously (Chow & Zhang, 2013; Guo et al., 2012). We only use the EMA data for the demonstrations in this book. Among the variables used for demonstrations are composite scores of positive affect and negative affect computed using item scores from the Positive Affect and Negative Affect Schedule (Watson, Clark, & Tellegen, 1988) and the circumplex model of affect (Larsen & Diener, 1992; Russell, 1980), measured on a scale of 1 (never) to 4 (very often). All items were assessed five times daily at partially randomized intervals that included both daytime assessments as well at least one assessment in the evening. The participants were asked to keep between an hour and a half and four hours between two successive assessments...irregularly spaced but data aggregation within equally spaced time blocks (midnight to noon; noon to midnight) was used to create an equidistant data set for illustration in Chapter 7.

1.3.4 fMRI Data

Data from two sources are used as examples of IAV analysis of functional MRI human brain data. The first comes from a publicly and freely available Autism Brain Imaging Data Exchange (ABIDE) repository and is used to demonstrate fundamental concepts in vector autoregression (Chapter 4). The individuals' data set used here is from a 24-year-old female (IQ = 146, ambidextrous) who has an autism spectrum disorder (ASD) diagnosis. The data are $T = 947$ sequential observations that were during a resting state block. Here, participants are instructed to not engage in any particular task and stare straight ahead at a crosshair. The second set of data comes from the lab of Kristen Lindquist. This set of data includes $N = 30$ participants who participated in a task meant to elicit anger and anxiety as well as a resting state block. Each of these three blocks were $T = 150$. The data have been used and described elsewhere (Lindquist et al., In Prep) and used here in Chapter 5 to demonstrate estimation with a large number of variables.

1.4 Notation

For the readers' ease, we have provided a list of commonly used notation and matrix definitions in the front matter of this book. Some principles are used throughout the book. For instance, matrices are denoted by uppercase bold letters; column vectors by lowercase bold letters; and scalars are italicized without bolding. No special notation is used to distinguish random (scalar, vector- or matrix-valued) variables from fixed variables, instead this distinction is made explicit in the definition of each variable. Latent variables are denoted by Greek letters, and manifest (observed) variables by Roman letters.

1.5 Conclusion

The present book aims to provide researchers with the skills, tools, and understanding needed to approach IAV via intensive longitudinal analysis. Rooted strongly within a time series perspective, the approaches presented here have been vetted for use with data commonly seen in the social sciences. Motivating examples using such data provide relevance for researchers looking for appropriate methods with which to attend to their research questions. The more quantitatively oriented reader will appreciate the proofs and details regarding technical aspects of the analytic. In the end, readers will understand the issues related to intensively gathered longitudinal data and how to accommodate them as well as conduct state-of-the-art analysis of these data.

References

Allport, G. W. (1937). *Personality: A psychological interpretation*. New York: Henry Holt.

Ballard, D.H., Hinton, G.E., & Sejnowski, T.J. (1983). Parallel visual computation. *Nature, 306*(5938), 21–26.

Bateson, G. (1972). *Steps to an ecology of mind: Collected essays in anthropology, psychiatry, evolution, and epistemology*. Chicago, IL: University of Chicago Press.

Bergman, L.R. & Magnusson, D. (1997). A person-oriented approach in research on developmental psychopathology. *Development and Psychopathology, 9*(2), 291–319.

Bollerslev, T. (1986). Generalized autoregressive conditional heteroskedasticity. *Journal of Econometrics, 31*(3), 307–327.

Borkenau, P. & Ostendorf, F. (1998). The Big Five as states: How useful is the five-factor model to describe intraindividual variations over time? *Journal of Research in Personality, 32*(2), 202–221.

Bouton, M.E. (2007). *Learning and behavior: A contemporary synthesis.* Sunderland, MA: Sinauer Associates.

Box, G.E.P. & Jenkins, G.M. (1970). *Time series analysis: Forecasting and control.* San Francisco, CA: Holden-Day.

Box, G.E.P. & Tiao, G.C. (1973). *Bayesian inference in statistical inference.* Reading, MA: Adison-Wesley.

Brillinger, D. (1975). *Time series data analysis and theory.* New York, NY: Holt Rinehart.

Cattell, R.B. (1946). *The description and measurement of personality.* New York, NY: New York World Book Company.

Chakraborty, B. & Moodie, E.E. (2013). *Statistical methods for dynamic treatment regimes* (pp. 31–52). New York, NY: Springer.

Chow, S.M. & Zhang, G. (2013). Nonlinear regime-switching state-space (RSSS) models. *Psychometrika, 78*(4), 740–768.

Conner, T.S., Tennen, H., Fleeson, W., & Barrett, L.F. (2009). Experience sampling methods: A modern idiographic approach to personality research. *Social and Personality Psychology Compass, 3*(3), 292–313.

Csikszentmihalyi, M. & Larson, R. (1987). Validity and reliability of the experience-sampling method. *Journal of Nervous & Mental Disease, 175*(9), 526–537.

Drabowitch, S., Papiernik, A., Griffiths, H., Encinas, J., & Smith, B.L. (2010). *Modern antennas.* Springer Science & Business Media. The Netherlands.

Durbin, J. & Koopman, S.J. (2012). *Time series analysis by state space methods* (Vol. 38). Cambridge: Oxford University Press.

Emotions and Dynamic Systems Laboratory (2010). The affective dynamics and individual differences (ADID) study: developing non-stationary and network-based methods for modeling the perception and physiology of emotions, Unpublished manual, University of North Carolina at Chapel Hill.

Finn, E.S., Shen, X., Scheinost, D., Rosenberg, M.D., Huang, J., Chun, M.M., & Constable, R.T. (2015). Functional connectome fingerprinting: Identifying individuals using patterns of brain connectivity. *Nature Neuroscience, 18*(11), 1664–1671. doi: 10.1038/nn.4135

Ford, D.H. & Lerner, R.M. (1992). *Developmental systems theory: An integrative approach.* Thousand Oaks, CA: Sage Publications, Inc.

Fortunato, S. (2010). Community detection in graphs. *Physics Reports, 486*(3–5), 75–174.

Goodwin, G.C. & Sin, K.S. (1984). *Adaptive filtering, prediction, and control.* Englewood Cliffs, N.J: Prentice Hall.

Gottlieb, G. (1992). *Individual development and evolution: The genesis of novel behavior.* New York, NY: Oxford University Press.

Gottlieb, G. (1995). Some conceptual deficiencies in developmental behavior genetics. *Human Development, 38*(3), 131–141.

Gottlieb, G. (2003). On making behavioral genetics truly developmental. *Human Development, 46*(6), 337–355.

Guo, R., Zhu, H., Chow, S.M., & Ibrahim, J.G. (2012). Bayesian lasso for semiparametric structural equation models. *Biometrics, 68*(2), 567–577.

Hamaker, E.L., Dolan, C.V., & Molenaar, P.C. (2005). Statistical modeling of the individual: Rationale and application of multivariate stationary time series analysis. *Multivariate Behavioral Research, 40*(2), 207–233.

Hannan, E.J. (1970). *Multiple time series.* New York, NY: Wiley.

Hannan, E.J. & Deistler, M. (1988). *The statistical theory of linear systems.* New York: Wiley.

Hansen, L.P. & Sargent, T.J. (2008). *Robustness.* Princeton, NJ: Princeton University Press.

Ho, M.-W. (2013). How development directs evolution. In P.C. Molenaar, R.M. Lerner, & K.M. Newell (Eds.), *Handbook of developmental systems theory and methodology.* New York, NY: Guilford Publications.

Holtzman, W.H. (1963). Statistical models for the study of change in the single case.*Problems in Measuring Change*, 199–211.

Hoyle, R.H. (Ed.). (2012). *Handbook of structural equation modeling.* New York, NY: Guilford Press.

Itô, K. (1944). 109. Stochastic integral. *Proceedings of the Imperial Academy, 20*(8), 519–524.

Jenkins, G.M. & Watts, D.G. (1968). *Spectral analysis and its applications.* San Francisco, CA: Holden-Day.

Jensen, A.R. (2006). *Clocking the mind: Mental chronometry and individual differences.* Elsevier. The Netherlands.

Johnston, T.D., & Lickliter, R. (2009). A developmental systems theory perspective on psychological change. In J.P. Spencer, M.S.C. Thomas, & J.L. McClelland (Eds.), *Toward a unified theory of development: Connectionism and dynamic systems theory re-considered* (pp. 285–296). New York, NY: Oxford University Press.

Klein, J.L. (1997). *Statistical visions in time: A history of time series analysis, 1662-1938.* Cambridge, MA: Cambridge University Press.

Kolmogorov, A.N. (1933). *Foundations of probability.* Oxford: Chelsea Publishing Co.

Kwon, W.H. & Han, S.H. (2005). *Receding horizon control: Model predictive control for state models.* London: Springer.

Kwon, W.H. & Han, S.H. (2005). *Receding horizon control: Model predictive control for state models.* London: Springer.

Larsen, R.J. & Diener, E. (1992). Promises and problems with the circumplex model of emotion. In M.S. Clark (Ed.), *Review of personality and social psychology* (Vol. 13). Newbury Park, CA: Sage.

Lazarsfeld, P.F. (1959). Latent structure analysis. In S. Koch (Ed.), *Psychology: A study of a science* (Vol. 3; pp. 476–543). New York: McGraw-Hill.

Lerner, R.M. (2002). *Concepts and theories of human development.* 3rd edition. Mahwah, NJ: Erlbaum.

Lévy, P. (1937). *Théorie de l'Addition des Variables Aleatoires.* Paris: Gauthier-Villars.

Li, S. C., Huxhold, O., & Schmiedek, F. (2004). Aging and attenuated processing robustness. *Gerontology, 50*(1), 28–34.

Lickliter, R. & Honeycutt, H. (2009). Rethinking epigenesis and evolution in light of developmental science. In *Handbook of behavioral and comparative neuroscience: Epigenetics, evolution, and behavior.*

Lickliter, R., & Honeycutt, H. (2010). Rethinking epigenesis and evolution in light of developmental science. In M. S. Blumberg, J. H. Freeman, & S. R. Robinson (Eds.), Oxford handbook of developmental behavioral neuroscience (pp. 30–47). Oxford University Press.

Lord, F.M. & Novick, M.R. (1968). *Statistical theories of mental test scores.* Reading, MA: Addison-Wesley.

Luce, M.F., Payne, J.W., & Bettman, J.R. (1999). Emotional trade-off difficulty and choice. *Journal of Marketing Research*, 36(2), 143–159.

Luce, R.D. (1999). *Utility of gains and losses: Measurement-theoretical and experimental approaches.* Mahwah, NJ: Lawrence Erlbaum.

Martin, V., Hurn, S., & Harris, D. (2012). *Econometric modelling with time series: Specification, estimation and testing.* Cambridge, MA: Cambridge University Press.

Miller, M.B., Van Horn, J., Wolford, G.L., Handy, T.C., Valsangkar-Smyth, M., Inati, S., Grafton, S., & Gazzaniga, M.S. (2002). Extensive individual differences in brain activations during episodic retrieval are reliable over time. *Journal of Cognitive Neuroscience, 14*(8), 1200–1214.

Molenaar, P. (2008). On the implications of the classical ergodic theorems: Analysis of developmental processes has to focus on intra-individual variation. *Developmental Psychobiology, 50*(1), 60–69.

Molenaar, P.C. & Campbell, C.G. (2009). The new person-specific paradigm in psychology. *Current Directions in Psychological Science, 18*(2), 112–117.

Molenaar, P.C.M., Lerner, R.M., & Newell, K.M. (2014). *Handbook of developmental systems theory & methodology.* New York: The Guilford Press.

Mulaik, S.A. (1987). A brief history of the philosophical foundations of exploratory factor analysis. *Multivariate Behavioral Research, 22*(3), 267–305.

Nesselroade, J.R. & Ford, D.H. (1985). P-technique comes of age multivariate, replicated, single-subject designs for research on older adults. *Research on Aging, 7*(1), 46–80.

Newell, K.M., Mayer-Kress, G., Hong, S.L., & Liu, Y.T. (2010). Decomposing the performance dynamics of learning through time scales. In P.C.M. Molenaar & K.M. Newell (Eds.), *Individual pathways of change: Statistical models for analyzing learning and development* (pp. 71–86). Washington DC: American Psychological Association.

Ostendorf, F. (1990). *Sprache und Persönlichkeitsstruktur. Zur Validität des Fünf-Faktoren-Modells der Persönlichkeit.* Regensburg: Roderer.

Overton, W.F. (2014). Relational developmental systems and developmental science: A focus on methodology. In P.C.M. Molenaar, R.M. Lerner, & K.M. Newell (Eds.), *Handbook of developmental systems theory & methodology* (pp. 19–65). New York: The Guilford Press.

Papoulis, A. (1977). *Signal analysis* (Vol. 191). New York, NY: McGraw-Hill.

Pierce, W.D. & Cheney, C.D. (2013). *Behavior analysis and learning.* London: Psychology Press.

R Core Team (2014). *R: A language and environment for statistical computing.* Vienna; R Foundation for Statistical Computing.

Robinson, E.A. & Silvia, M.T. (1981). *Digital foundations of time series analysis, Vol. 2: Wave-equation space-time processing.* San Francisco: Holden-Day.

Robinson, O.C. (2011). The idiographic/nomothetic dichotomy: Tracing historical origins of contemporary confusions. *History and Philosophy of Psychology, 13*(2), 32–39.

Russell, J.A. (1980). A circumplex model of affect. *Journal of Personality and Social Psychology, 39*(6), 1161–1178.

Saunders, P.T. (2013). Dynmical systems, the epigenetic landscape, and punctuated equilibria. Development directs evolution. In P.C. Molenaar, R.M. Lerner, & K.M. Newell (Eds.), *Handbook of developmental systems theory and methodology.* New York, NY: Guilford Publications.

Schwartz, D. & Collins, F. (2007). Environmental biology and human disease. *Science, 316*(5825), 695–696.

Smith, L.B. & Thelen, E. (2003). Development as a dynamic system. *Trends in Cognitive Sciences, 7*(8), 343–348.

Stone, A. A., & Shiffman, S. (1994). Ecological momentary assessment (EMA) in behavorial medicine. *Annals of Behavioral Medicine*, 16(3), 199–202. https://doi.org/10.1093/abm/16.3.199

Stratonovich, R.L. (1966). A new representation for stochastic integrals and equations. *SIAM Journal on Control*, 4(2), 362–371.

Townsend, J.T. & Ashby, F.G. (1984). Measurement scales and statistics: The misconception misconceived. *Psychological Bulletin*, *96*(2), 394–401.

Trull, T. J. & Ebner-Priemer, U. W. (2009). Using experience sampling methods/ecological momentary assessment (ESM/EMA) in clinical assessment and clinical research: Introduction to the special section. *Psychological Assessment*, Dec; 21(4): 457–462.

Watson, D., Clark, L.A., & Tellegen, A. (1988). Development and validation of brief measures of positive and negative affect: The PANAS scales. *Journal of Personality and Social Psychology*, *54*(6), 1063.

Whittle, P. (1990). A risk-sensitive maximum principle. *Systems & Control Letters*, *15*(3), 183–192.

Wiener, N. (1930). Generalized harmonic analysis. *Acta Mathematica*, *55*(1), 117–258.

Windelband, W. (1901). *A history of philosophy; with special reference to the formation and development of its problems and conceptions*. New York, NY: Macmillan.

Wohlwill, J.F. (1973). The concept of experience: S or R? *Human Development*, *16*(1–2), 90–107.

Wold, H. (1938). *A study in the analysis of stationary time series*. Stockholm: Almqvist & Wiksell.

Yule, G.U. (1926). Why do we sometimes get nonsense-correlations between time-series? A study in sampling and the nature of time-series. *Journal of the Royal Statistical Society*, *89*(1), 1–63.

Appendix: Heuristic Introduction to Time Series Analysis for Psychologists

Suppose that in terms of Cattell's box (Figure 1.1), we select a single variable and assess the scores of a given person P at T equidistant occasions or time points $t = 1,2, \ldots, T$. The univariate time series thus obtained can be denoted by: y_{Pt}, $t = 1,2, \ldots, T$. However, the P-subscript is redundant and therefore will be omitted. Hence, the observed univariate time series for P is denoted by: y_t, $t = 1,2, \ldots, T$. Later on a slightly different notation is used.

In what follows the denotation "time series" will be understood as "single-subject time series" unless a different interpretation is explicitly mentioned. In order to better appreciate the concept of time series, we turn to Chapter 2 of Lord and Novick's (1968) classic book on test theory:

> *Let us suppose that we repeatedly administer a given test to a subject and thus obtain a measurement each day for a number of days. Further, let us assume that with respect to the particular trait the test is designed to measure, the person does*

> *not change from day to day and that successive measurements are unaffected by previous measurements. Changes in the environment or the state of the person typically result in some day-to-day variation which is obtained.*
>
> (Lord & Novick, 1968, pp. 27–28)

Some key observations can be extracted from this quotation. Firstly, repeated administration of the same test typically will yield a time series that shows day-to-day variation. This variation is IAV (within the fixed Person P). Secondly, it is assumed that the trait score of P which the test is designed to measure is constant across days. Thirdly, it is assumed that repeated measurement of P across days does not affect P.

Lord and Novick cite Lazarsfeld (1959) in order to further explain the third point mentioned above, namely, that it is assumed that repeated measurement does not affect P:

> *Suppose we ask an individual, Mr. Brown, repeatedly whether he is in favour of the United Nations; suppose further that after each question we "wash his brains" and ask him the same question again. Because Mr. Brown is not certain as to how he feels about the United Nations, he will give sometimes a favourable and sometimes an unfavourable answer. Having gone through this procedure many times, we then compute the proportion of times Mr. Brown was in favour of the United Nations …*

The "brain washing" of Mr. Brown is meant to remove any memories about, and other after-effects of, the previous assessment of his opinion about the United Nations. Accordingly, he answers the question at each measurement occasion concerned as if it was posed to him for the first time. This is the interpretation of the third point mentioned above, namely that it is assumed that repeated measurement does not affect P. Lord and Novick (1968) further elaborate this interpretation as follows:

> *Most students taking college entrance examinations are convinced that how they do on a particular day depends to some extent on "how they feel that day". A student who receives scores which he considers surprisingly low often attributes this unfortunate circumstance to a physical or psychological indisposition or to some serious temporary state of affairs not related to the fact that he is taking the test that day. To provide a mathematical model for such cyclic variations, we conceive initially of a sequence of independent observations, as described by Lazarsfeld, and consider some effects, such as the subject's ability, to be constant, and others, such as the transient state of the person, to be random.*
>
> (Lord & Novick, 1968, p. 30)

The general picture emerging from these citations is as follows. Repeated measurement of a single fixed person P with a given test yields a univariate time series of scores. The average of these scores is indicative of the trait score of P, whereas the variation about this average level is random. Using the notation introduced above, we denote the univariate time series of scores by y_t, $t = 1,2, \ldots, T$. Then, the average of these scores (the level of the time series) is estimated by

$$m = T^{-1} \sum_{t=1,T} y_t$$

We can interpret m given as the estimated trait score of P. Notice that it is simply the average of P's scores across the measurement occasions. Hence, the time points $t = 1,2, \ldots, T$ fulfil the same role as the subjects in an analysis of IEV: the time points constitute the "cases" or "replications" over which pool occurs in order to compute an estimate (in this case an estimate of the mean level).

Because of the "brain washing" that is supposed to remove all carry-over effects between consecutive measurements, the centered time series y_t − m, $t = 1,2, \ldots, T$, is characterized by Lord and Novick as "a sequence of independent observations". That implies that the time series can be conceived of as a sequence of observations that are mutually independent, in a similar vein as the observations obtained with different subjects in an analysis of IEV are mutually independent. Of course, the "brain washing" described by Lazarsfeld is impossible to carry out; it is an idealization. Lord and Novick (1968) remark:

> *In mental testing we can perhaps repeat a measurement once or twice, but if we attempt further repetitions, the examinee's responses change substantially because of fatigue or practice effects.*
>
> (Lord & Novick, 1968, p. 13)

This implies that in reality a time series of observed scores of a fixed Person P will not constitute "a sequence of independent observations". Instead, the time series will show sequential dependencies, for instance, due to the fatigue or practice effects mentioned by Lord and Novick. Clearly the latter fatigue and practice effects can be considered to be contaminating effects, i.e., aspects of the time series of P's mental test scores in which we are not primarily interested. But, in general, the presence of sequential dependencies has a much more fundamental status in that they provide the prime source of information about the process in which we are interested. To explain why sequential dependencies are fundamental in time series analysis requires the availability of formal concepts, some of which will be introduced shortly. To start with some heuristic explanation are given.

When studying the IAV of a psychological characteristic of a single subject P, we can conceive of the time series thus obtained as being generated by a dynamic system. For instance, a time series of mental test scores can be regarded as being generated by P's information processing system. Also, a time series of electro-encephalographic values can be conceived of as being generated by P's brain system; a time series of anxiety scores can be conceived of as being generated by P's emotional system; a time series of personality scores can be conceived of as being generated by P's personality system, and so forth. According to this perspective, P can be regarded as a high-dimensional dynamic system, the time-dependent behavior of which can be measured in the form of an equally high-dimensional time series. In reality, of course, we will have to focus on measuring the time-dependent variation of only a subset of P's psychological characteristics, where each characteristic corresponds to a selected variable in terms of Cattell's Data Box (cf. Figure 1.1).

Each real dynamic system has a momentum, i.e., a tendency to persist in its momentary state of activity. For physical systems in motion, this is a direct consequence of Newton's laws—think of the difficulty to stop a speeding car in case of emergency. But no real dynamic system has the capacity to alter its state instantaneously in arbitrary ways—that would necessitate the expenditure of potentially infinitely large pulses of energy. The same holds for psychological processes; for instance, it is not possible to instantaneously stop laughing, change one's mood, lower one's anxiety level, or start sleeping.

Not only has each dynamic system a tendency to persist in its momentary state of activity, but the rate of change of its behavior also has a characteristic time scale. The activity of nervous systems changes on a typical time scale of milliseconds, whereas learning processes take place on a much slower time scale. It therefore makes sense to assign to each dynamic system a so-called "natural" time scale characterizing its rate of change. In general, a dynamic system's momentum will be linked to its natural time scale. If a dynamic system is repeatedly measured with a sampling frequency that is accommodated to its natural time scale, the system's momentum will give rise to sequential dependencies in the time series thus obtained. These sequential dependencies reflect the dynamic regularities characterizing the system's behavior and can be used to predict its future behavior (as well as retrodict its previous behavior and/or interpolate within the observation interval). Therefore, the analysis of sequential dependencies is fundamental in time series analysis.

2

Ergodic Theory: Mathematical Theorems about the Relation between Analysis of IAV and IEV

2.1 Introduction

Until now the standard approach to quantitative analysis of psychological data has been based on the analysis of IEV. It consists in its simplest form of the following steps: (a) draw a sample from a homogeneous population of interest, meaning that different individuals sampled from the population can be regarded as independent and inter-changeable replicates of each other; (b) compute statistics of interest by pooling across the assessments of this sample (the hallmark of analysis of IEV); (c) generalize the results thus obtained to the population from which the sample was drawn. These results are valid at the level of the population. Yet, because the population is considered homogeneous it is also assumed that they pertain to each individual making up this population. The latter assumption implies that results that are obtained by means of the analysis of IEV, if valid at the population level, would also apply at the individual level (which is characterized by IAV). In this chapter, we will scrutinize this generalization from the population to the individual level of results obtained in the analysis of IEV.

We encountered an instance of this generalization in the previous chapter while discussing the classical test theory (CTT). There it was explained that the definition of the true score is in terms of the mean of the distribution characterizing a given, fixed individual, where this distribution is derived from an individual time series, namely, time series of repeated, presumably independent, assessments of this individual. Hence, the distribution is based on the analysis of IAV. Because it was deemed impossible to obtain a sufficient number of repeated assessments which are independent, the perspective was shifted from a large number of repeated assessments of a given individual to one (or a few repeated) assessment(s) of a large number of individuals. The latter involves analysis of IEV. Yet CTT produces results (tests) that often are used to obtain individual assessments, hence involving generalization

DOI: 10.1201/9780429172649-2

of results valid at the population level to the level of individuals. In particular, a test's reliability is valid at the population level but also determines the standard error of individual estimates of true scores. Lord and Novick (1968) prove that the decomposition of IEV error variance, *observed variance* = *true variance* + *error variance*, where *true variance* in this particular context is the mean of the variances of intra-individual distributions. Insofar as the variances of intra-individual distributions can differ arbitrarily across individuals (i.e., are heteroscedastic), the assumption that the same value of *error variance* holds for all individuals will be inappropriate (see Molenaar, 2008, for further elaboration).

In this chapter, the focus is on the validity of generalizing results obtained in the analysis of IEV to the level of individuals and, vice versa, generalizing results obtained in the analysis of IAV to the population level. If they are generalizable in this way, one can say the process is *ergodic*—a concept to be defined in detail below. That is, the qualities of a process when looking across individuals are the same as when looking across time for one of the individuals. When considering just the mean and standard deviation, an ergodic system would imply that every individual has the same expected values. Figure 2.1 provides a depiction of an ergodic sample. Here, the data were randomly generated from a normal distribution with a mean of 5 and a standard deviation of 1 for each individual with no dependencies across time. At any point in time, the IEV average across the $N = 300$ individuals is nearly the same as the IAV average obtained when looking across the $T = 300$ time points for each individual.

Although not by name, the concept of ergodicity has been discussed before in the history of quantitative psychology, and in the next section, a brief historic overview is given. We then provide some foundational information that is necessary for understanding the concept of ergodicity from a mathematical perspective as well as other analytic and inferential approaches discussed throughout this book. Then, a definitive answer to the ongoing debate about the relation between results obtained in analyses of IEV and IAV is provided for, specifying when, and under what conditions, ergodicity can be assumed in Gaussian processes, while making use of a general theorem of classic ergodic theory (Molenaar, 2004). Some immediate consequences of this definitive result are discussed.

2.2 Some History Regarding Generalizability of IEV and IAV Results

In the early behavioristic era of psychology, the question about the relation between the average learning curve (at the population level) and the individual learning curves on which this average is based was discussed.

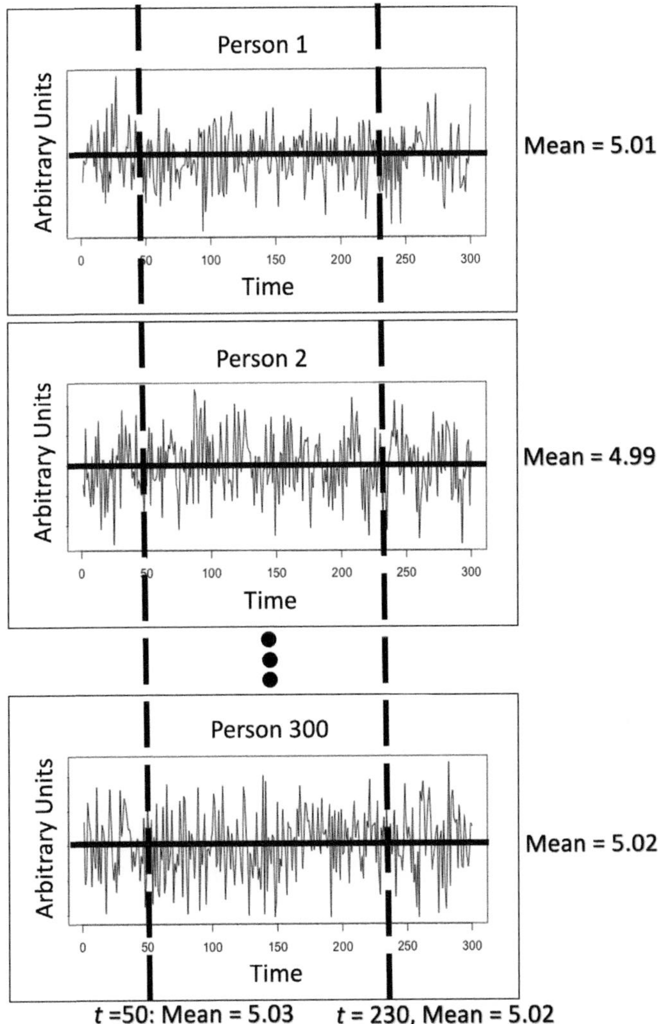

FIGURE 2.1
Depiction of an ergodic process. The average across ensembles (here, $N = 300$ participants) is nearly the same as the average across time ($T = 300$ for each individual).

The solid lines represent looking across "time" and IAV. The dashed lines indicate looking across "people" and are comparable to IEV.

Sidman (1960) refers to several papers on this topic, including Merrill (1931), Sidman (1952), Hayes (1953), Bakan (1954), and Estes (1956). A comprehensive treatment of this question was given by Hannan (1991), who concluded that a direct mapping between the average learning curve and individual learning curves is only possible if the functional forms of the individual learning curves are linear.

Lamiell (1987, 2003) has persistently criticized the dominant use of IEV-based analysis in the study of personality. His main point of view is that differences between persons do not convey essential information about the personality system characterizing each individual person. Within developmental psychology, Developmental systems theory (DST) has been critical of IEV-based approaches (see, for instance, Lerner & Schmid, 2013). DST emphasizes that each individual develops in relation to a complex set of potentially unique ecological conditions, requiring analysis of IAV as the basic approach. Another important proponent of DST is Gottlieb (1997, 1998), who criticizes the IEV-based analysis underlying developmental behavior genetics.

Deserving of a special mention is the lifelong work of John Nesselroade, a student of Cattell, on the application of P-technique (factor analysis of individual-level time series data discussed in the next chapter). His paper, co-authored by Don Ford (Nesselroade & Ford, 1985), presents multivariate replicated single-subject designs as the optimal approach to study inter-individual differences in intra-individual change (the focus of life span psychology). Also important is the generalization of tests for measurement invariance to the level of latent variables (Idiographic Filter, IF) under the assumption that different individuals assign different meanings to observed variables (Nesselroade et al., 2007). The IF assumes the covariance among latent constructs is homogeneous across individuals, but only after idiosyncrasies in the measurement process—namely, in how latent variables and factors are linked to observed variables—have been accounted for. The IF was originally based on P-technique factor analysis, but was extended to state-space models in Molenaar and Nesselroade (2012) to enable lagged relations among latent constructs.

Increasingly scientists across multiple domains of social inquiry have noted the discrepancy between IAV and IEV results. Psychophysiological studies regularly assess the dynamic attributes of individuals. For instance, aspects of the heart rate across time are known to vary meaningfully across individuals (Porges, 2007), precluding the ability to develop one model that describes all individuals' variability across time. In neuroimaging studies, the between-subject findings regarding brain processes do not always provide reliable models for within-subject brain activity. As such, individual-level modeling has become a priority for neuroimagers (Finn et al., 2015; Smith, 2012). Daily diary studies do not always have the data required for complete individual-level analysis, but emerging evidence supports differentiation in temporal processes. For instance, at the sample level, individuals with borderline personality disorder might have higher reports of impulsivity and moodiness. However, across time, not all individuals with Borderline personality disorder diagnoses have the same covariation of symptoms (Wright, Hopwood, & Simms, 2015). In fact, some claim that multiple temporal processes can give rise to the same diagnostic category (Fair et al., 2012; Gottesman & Gould, 2003; Volkmar et al., 2004), and thus investigation into individual-level models is absolutely necessary to capture the varied patterns as no *one* model will explain all (or even the majority of) individuals. These examples provide substantive motivation for investigating human processes across time rather than solely across individuals.

This concludes our admittedly concise history of discussions about the relation between results obtained in the analysis of IEV with those obtained in the analysis of IAV. These discussions were theoretical, presenting arguments in words, and therefore inconclusive. In 2004, Molenaar presented a definitive answer to the question about the relation between results of analysis of IEV versus IAV, making use of classical ergodic theory. Before providing a heuristic introduction to this theory, we first present some fundamental concepts required to elucidate the ergodic theorem.

2.3 Two Conceptualizations of Time Series

The general formal definition of a time series is as follows: Multivariate time series data is a set of ordered random variables indexed by time.[1] For explanatory purposes attention is restricted here to ordered sets of univariate random variables, i.e., univariate time series.

An abstract expression for a univariate time series then is

$$\{y(\omega,t), t = 0, \pm 1, \pm 2, \ldots\}. \tag{2.1}$$

Here, y is the measured variable across time (e.g., a mental test, heart rate); $\omega \in \Omega$ is the possible value of an unobserved random process (Ω is the space of all possible values of measurement for process); and t is the time index. Notice time is supposed to extend from minus infinity to plus infinity. Time is taken to be discrete; it is represented as evolving in discrete steps of equal length of one unit. The latter convention—that the interval between consecutive time points is one unit—is chosen to ease notation (a more general convention would be that this interval has length Δt which can take any value). The assumption that each pair of consecutive time points is separated by equidistant intervals is standard in elementary time series analysis. Replacement of this assumption by a more general convention according to which intervals between pairs of consecutive time points may vary (or even may themselves be random) leads to a much more complicated analysis. Of course, time also could be represented as a continuous variable. But this too leads to a more complicated analysis that will be considered later on in Chapter 7.

There are basically two ways in which the abstract expression for a time series can be interpreted. Each interpretation focuses on one of the two arguments of $y(\omega, t)$, i.e., t or ω. According to the first interpretation (which goes back to Kolmogorov, 1933), at each time point t, the time series has a random

[1] The index can be generalized to space-time, but this generalization will not be considered in this section.

distribution (i.e., probability mass function) as expressed by its dependence on $\omega \in \Omega$, the possible values of an unobserved random process. This interpretation leads to the concept of finite-dimensional distributions characterizing a time series. That is, at a given set of consecutive T time points, $t = 1, 2, \ldots, T$, the time series for each ω value provides a T-dimensional random distribution. Each value ω has a distinct probability of being the measurement value at each time point.

According to the second interpretation (which goes back to Wiener, 1930), each outcome value of the random process, ω, is associated with a time series that takes on different values as a univariate function of time (as expressed by its dependence on $t = 0, \pm 1, \pm 2, \ldots$). This interpretation leads to the concept of an ensemble—or collection—of random time-dependent functions, indexed by ω, where $P(\omega, t)$ denotes the density of each possible ω value in the ensemble over time. In contrast to Kolmogorov's description which focuses on the time series for each potential ω value, Wiener's interpretation focuses on the probability function indexed by each t. Thus, the probability for each ω value is a function of time for Wiener. This contrasts Kolmogorov's description, wherein the distribution of ω values is considered across all time points.

Both interpretations converge to equivalent conceptual foundations of time series analysis (see Brillinger, 1975). Therefore, we make free use of both interpretations in what follows. That is, *we will conceive of a time series both as an ordered set of finite-dimensional random distributions and as a random function of time*. Next, we define some foundational concepts critical for understanding the concept of ergodicity, including first and second moments, and stationarity. Readers already familiar with these concepts may skip ahead directly to Section 2.5.

2.4 Some Preliminaries

Much like in cross-sectional research, the first-order moment of a random variable across time is the mean of its distribution. If the data are multivariate, the first-order moment is the mean vector of its multivariate distribution. The first-order moment of the random variable y is denoted as $E[y]$, shorthand for the "Expectation of y". The second-order central moment of a random variable is its variance. Below, we use the notation $\text{cov}[y,y']$ to denote the variance function of y, where y' indicates the transpose of y. If the random variable is vector-valued at each time point (as in the multivariate case), the second-order central moment is the covariance matrix of its multivariate distribution.

The first-order moment associated with the univariate time series is

$$E[y(\omega,t)] = \mu(t), \forall t, \tag{2.2}$$

where it is understood that the expectation is taken with respect to the density, P(ω, t)—namely, as average over all possible values of ω. The notation $\forall t$ is shorthand for: for all time points $t = 0, \pm 1, \pm 2, \ldots$. Hence, the first-order moment of a time series is a function of time: it is the first-order moment function (of time) $\mu(t)$. We will refer to $\mu(t)$ as the *mean function*, an average over all possible values of ω, for a particular time point.

Equation (2.2) is an expression of the fact that the time series $y(\omega,t)$ defines a random distribution at each time point, t, the mean of which can be different from the means at all other time points. Hence, the mean function of the time series is allowed to be an arbitrary function of time. For practical applications, this arbitrariness is unsuitable, as will be apparent on a moment's reflection. In the typical application of time series analysis, one only has available a single time series, observed during a finite interval $t = 1, 2, \ldots, T$. Let us denote such an *observed time series* as

$$y(t), t = 1, 2, \ldots, T. \tag{2.3}$$

An observed time series defined by (2.3) consists of a *single* observation of the value of y at each time point t. The mean $\mu(t)$ (and all other moments) cannot properly be estimated on the basis of a single observation. The availability of other observed values of y at neighboring time points is of no use, because according to (2.2), the means at these other time points can differ arbitrarily from the mean at t. Therefore, the standard approach is to assume that the mean function is constant in time:

$$\mathrm{E}[y(\omega, t)] = \mu, \forall t \tag{2.4}$$

The assumption expressed by (2.4) is called: *stationarity of the mean function*. This is reflected in the lack of a time indicator for the mean μ. In Chapter 7, advanced analytic methods are introduced where such estimates and coefficients can change across time. For the majority of the book, we focus on the case where estimates are assumed to be constant as many modeling approaches require this assumption.

In comparison to (2.2), where the mean function is allowed to vary arbitrarily in time, the stationarity assumption (2.4) would seem to constitute an extreme restriction. One could think of less restrictive assumptions such as the assumption that the mean function obeys a finite-order polynomial in time. For instance, perhaps it has a quadratic form. One has to remember, however, that according to (2.1), a time series is defined for all times ranging from the arbitrarily distant past to the arbitrarily distant future. Many parametric models for time-varying mean functions quickly run off to infinity if the time variable approaches infinity, in particular polynomial models in time. This is not realistic as few processes will ever approach infinity. Hence, parametric models of the mean function of a time series have to obey severe restrictions in order to remain bounded for all possible times. For reasons such as these, elementary time series analysis is not primarily focused on the mean function and it is usually assumed that the mean function is stationary, as expressed by (2.4). Rather, time series analysis often focuses on the lagged effects, or how observations at previous time points can predict values at later time points. Henceforth it is assumed that each time series has a stationary mean function. If an observed time series displays some sort of time-varying change in mean, then for now we assume that this trend has been removed in a preliminary trend analysis (discussed in Chapter 4).

We now turn attention to what constitutes the main focus of elementary time series analysis: the central second-order moment function, which provides insight into lagged relations. The qualification "central" refers to the fact that the second-order moment expresses the covariance of centered variables, i.e., variables with their means subtracted. To start with, a formal definition of the second-order moment function is given, after which we will clarify other specifics. The central second-order moment function associated with time series (2.1) at times t and s is formally defined for a univariate series as

$$\sigma(t,s) = \text{cov}[(y(\omega,t) - \mu),(y(\omega,s) - \mu)], \forall t, \forall s. \tag{2.5}$$

Equation (2.5) expresses the covariance of the univariate time series $y(\omega,t)$ at time t and the same time series at time s: $y(\omega,s)$. It thus expresses the covariance of the time series with itself; therefore (2.5) is called the *autocovariance function*. It is a function of a pair of time points, t and s. In the case where $t = s$, $\sigma(t,t)$ is the *variance* of the time series at time point t.

As noted previously our standard assumption, unless indicated otherwise, is that time series have stationary mean functions. Therefore, the mean function can be taken to equal zero without affecting generality. That is, in the case where the series has a mean unequal to zero, we consider only the centered series. With this convention in hand (2.5) can immediately be simplified: $\sigma(t,s) = \text{cov}[y(\omega,t), y(\omega,s)]$, $\forall t$, $\forall s$. Notice that in (2.5), the apostrophe denoting the transpose of $y(\omega,s)$ has been omitted, because in this section we only consider univariate time series. The qualification "central" often will be omitted.

The autocovariance function defined by (2.5) is allowed to obtain different values for each different pair of time points. Given an observed univariate time series y (2.3), it is clear that no reasonable estimate of (2.5) can be obtained. This is because there is only a single observed value of $y(t)$ at each time $t = 1, 2, \ldots, T$. In order to be able to estimate the autocovariance function based on a single observed time series, we will have to constrain the possible forms of (2.5). Much like we did for the mean function, we can constrain the autocovariance function to not depend on time. The standard restrictive assumption in elementary time series analysis is the following. Let, without lack of generality, $s = t - u$, where $u = 0, \pm 1, \pm 2, \ldots$. Define the covariance function as

$$\sigma(t,t-u) = \sigma(s,s-u) = \sigma(0,u) = \sigma(u), \forall t, \forall s, \forall u. \tag{2.6}$$

According to (2.6), the autocovariance function of $y(\omega,s)$ does not depend upon absolute time, but only upon the relative time difference between pairs of time points t and s. This relative time difference is called the *lag*, and it is expressed by the values $u = 0, \pm 1, \pm 2, \ldots$. It is obvious from the definitions of (2.5) and (2.6) that the autocovariance function is *symmetrical* about lag $u = 0$: $\sigma(u) = \sigma(-u), \forall u$.

When the autocovariance function obeys (2.6), it is called stationary. Hence, a *stationary autocovariance function* is not a function of pairs of absolute time points, but rather the lag u between such pairs of time points. More specifically, (2.6) expresses the assumption that the autocovariance between time points $t = 1$ and $s = 2$ equals the autocovariance between the time points $t = 2$ and $s = 3$, etc. The autocovariance between all pairs of consecutive time points (separated by a lag of $u = 1$) equals $\sigma(1)$. Also, the autocovariance between all pairs of time points separated by a lag of $u = 2$ equals $\sigma(2)$, and so forth.

This brings us to a general term used to describe that both mean and autocovariance functions are constant across time: "weak stationarity". This assumption for many analytic approaches described in the present book (as well as time series literature in general) necessitates that the mean function is constant and the covariance function depends only on the lag u, the interval or time separation between any two occasions, and not each occasion's point in time. These conditions are often collectively referred to as simply, "stationarity", and data that conform to these assumptions are considered to be "stationary". For Gaussian processes, over-time invariance in mean and covariance functions already yields strict stationarity; for non-Gaussian processes with higher moments to consider, strict stationarity requires over-time invariance in all moments. Invariance in the first two moments only leads to so-called weak stationarity or second-order stationarity (Shumway & Stoffer, 2000).

The assumptions of stationarity of the mean function (2.4) and stationarity of the autocovariance function (2.6) together make possible the estimation of the autocovariance function based on a single observed time series (2.3). Taken the mean to be zero, the estimator of $\sigma(u)$ will be denoted by $c(u)$ and is defined as

$$c(u) = T^{-1} \sum_{t=u}^{T} \left[y(t) y(t-u) \right],\ u = 0, 1, \ldots, m \tag{2.7}$$

Notice that for $u = 0$, $c(0)$ is the estimate of the variance of the time series. Notice also that for $u \neq 0$ the number of cross-products $y(t)y(t - u)$ in (2.7) is $T - u$, whereas the sum of these cross-products is divided by T, not $T - u$. This yields a biased estimate of $\sigma(u)$ in case $u > 0$, and an unbiased analogue is obtained by replacing T^{-1} by $(T - u)^{-1}$. However, it can be proven (cf. Jenkins & Watts, 1968) that the standard error of (2.7) is much smaller than for its unbiased analogue. Also, it can be proven that using (2.7) guarantees positive-definiteness of the associated autocovariance matrix (cf. Box & Jenkins, 1970). Positive-definite matrices have a determinant larger than zero and positive eigenvalues, which are important features for estimation.

For an observed time series of length T, the autocovariance function is estimated up to maximum lag m, where m is strictly less than T: $m < T$. Usually m is taken much smaller than T in order to guarantee that $c(m)$, the autocovariance at maximum lag m, still is based on a sufficiently large number $T - m$ of cross-products in (2.7). With this foundational material in mind, we now turn to ergodic theory.

2.5 When Is a System Ergodic?

Birkhoff (1931) shows that the analysis across time and the analysis across space (i.e, people) only yield the same results if the system is ergodic (see the gray box for technical details, which can be skipped without loss of understanding the concept). Considering humans and keeping with the simple example of an average, we would expect a person's average on some behavior across time to equal the average seen in a cross-sectional study of others in their population. Hence, the expected average across time (obtained via IAV) is the same across space (i.e., individuals; obtained via IEV) if ergodicity holds. Earlier in this chapter, we saw this depicted in Figure 2.1. This concept can be extended to other parameters of interest in addition to the mean.

Proving that a given dynamical system is ergodic turns out to be exceedingly difficult and has been accomplished for only a few dynamical systems. But if the system is Gaussian, i.e., if its finite-dimensional distributions are multivariate normal, then the criteria for such a system to be ergodic are straightforward. The following criteria for Gaussian systems to be ergodic are given in Hannan (1970).

Necessary and sufficient criteria for a Gaussian system to be ergodic are as follows:

a. The system has to be stationary, that is, its mean function has to be constant in time and its covariance function has to be a function of lag only.
b. Different persons in the population are all characterized by replicates of the same dynamical system.

According to criterion (a), the Gaussian system should not have any time-varying (e.g., cyclical) trends, its variance should be invariant in time and the covariance between any two time points and $t = 1$ and $t = 2$ only should depend upon the relative difference (lag) $u = 2 - 1 = 1$. According to criterion (b), each person in the population should obey the same dynamical rules (same general form of mapping function or model, same model parameter values, etc.). If any (or both) of these criteria are violated then the system is non-ergodic and there no longer may exist a relation between results obtained in analysis of IEV with those obtained in analysis of IAV.

There is an immediate consequence of these criteria for the analysis of learning and developmental processes. These processes almost by definition have time-varying statistical characteristics such as changing means, variances and/or lagged covariances, and therefore violate criterion (a). Hence, learning and developmental processes are immediately considered non-ergodic and to obtain results for such processes that apply at the individual level it is required to carry out dedicated analysis of IAV. We now turn to the second and common cause for nonergodicity: heterogeneity.

2.6 Birkhoff's Theorem of Ergodicity

The term "ergodic" is due to Ludwig Boltzmann, the founding father of statistical mechanics (Boltzmann, 1885). He introduced it in his attempt to explain thermodynamic phenomena by means of mechanical models and their underlying mathematical principles. Botlzmann considered an ideal gas contained in a box and represented by d frictionless moving particles. Each particle is described by six coordinates (three for position and three for momentum), so properties of the gas (i.e., the state of the system) at any point $x \in \Re^{6d}$ are summarized by 6d real values. Clearly, not all points in $\Re^{6d}$ can be attained by the gas in the box, so consideration is restricted to the set X of all possible states and this set is called the state space of the system. It is observed that the system changes while time is running, i.e., the particles are moving (in the box) and therefore a given state (point in X) also changes (in X) across time. This motion is governed by Hamilton's equations. The solutions to these equations define a mapping function, denoted with T (not to be confused with our prior use of T = length of observations) which transforms a given state to another state that has the same parameterization (e.g., expected mean). This is expressed mathematically as $T\colon X \rightarrow X$. Specifically, if the system is in state $x_0 \in X$ at time $t = 0$, then at time $t = 1$, it will be in a new state $x_1 = T(x_0)$. As a consequence, the state at time $t = 2$ becomes $x_2 = T^2(x_0)$ and

$$x_n = T^n(x_0), n = 1, \ldots, N \tag{2.8}$$

N can be a large integer (even taken to infinity). The so-obtained set $\{T^n(x_0) | n\}$ is called the *orbit* of x_0. In this way, the physical motion of the particles becomes the motion of the points in state space. This leads to the following definition:

Definition: A pair (X; T) consisting of a state space X and a map T: $X \rightarrow X$ is called a dynamical system. The dynamics may be deterministic: that is, given knowledge of its previous state and the mapping function, its future state can be predicted perfectly. The dynamics can also be probabilistic or stochastic, wherein its future state cannot be predicted perfectly even if perfect knowledge of its previous state and the mapping function.

Some dynamical systems are *measure preserving*, meaning that the measure (for example, the mean) of a set of points is always the same as the measure of the set from which it is mapped. In mathematical analysis, a measure on a set is a systematic way to assign a non-negative number to each suitable subset of that set, intuitively interpreted as its size. A probability measure assigns the value 1 to the set. In symbols, using P for the probability measure: $P(T^{-1}(A)) = P(A)$ for any measurable subset $A \in X$ of a measure-preserving dynamic system. From this, it follows that all nonzero invariant sets of an ergodic system have the same measure as the complete state space X. This leads to one important implication: all possible subsets of an ergodic system, provided that the subsets are large enough in size, have the same properties as the complete state space X for inferential purposes.

For ergodic systems, Birkhoff (1931) proved an important theorem. For an accessible presentation of the proof see Lindgren, 2013, Chapter 6. Let $f(x)$ denote a well-behaved (integrable) function of the space X. Then consider a randomly chosen point $x \in X$ and construct the asymptotic time average of $f(x)$ along the orbit of x:

$$\mathrm{f}^* = \lim_{n\to\infty} \frac{1}{n} \sum_{k=0}^{n} f[T^k(x)] \tag{2.9}$$

Birkhoff proved that this limit exists and is equal to the space average:

$$\mathrm{f}^{**} = \int_{-\infty}^{\infty} f(x)P(dx) \tag{2.10}$$

That is, for ergodic systems: f* = f** as the number of observations go to infinity.

To understand what this means, consider for $f(x)$ the indicator function $I_A(x)$ which equals 1 for $x \in A$ and equals zero otherwise. Then, (2.9) defines the proportion of times that the orbit hits the set $A \subset X$. The space average (2.10) gives the probability that a randomly chosen point x is a member of A. Both quantities reflect the size of A and are therefore (asymptotically) equal for ergodic processes. One may regard the asymptotic time average (2.9) as an analogue for the analysis of IAV, while the space average (2.10) corresponds to the analysis of IEV. In the next section, a straightforward test of ergodicity is presented.

2.7 Heterogeneity as Cause of Non-Ergodicity

In Molenaar (2004), a reference is made to a number of published simulation studies showing that analysis of IEV is insensitive to the existence of large-scale violations of criterion (b) for ergodicity: that different persons in a population have to obey the same dynamical model. Cross-sectional, applied and behavior genetic factor models in which each replication (participant) was characterized by its own model (participant-specific number of factors, factor loadings, and/or measurement error variances) were used to simulate heterogeneous data sets violating criterion (b). Application of standard IEV factor models to these data yielded satisfactorily fitting models even though this cross-sectional model failed to describe the majority of individuals. Hence, the assumption of participant-invariant parametric models, which is essential in obtaining the (maximum likelihood) model fits, was violated extensively. A formal proof of this insensitivity of analysis of IEV to violations of criterion (b) is given in Kelderman and Molenaar (2007).

There exists mathematical-biological evidence for the presence of widespread heterogeneity (violating criterion (b) for ergodicity) in human populations. This is based on the phenomenon of biological pattern formation which explains how an initially homogeneous tissue develops into a differentiated system. So-called reaction-diffusion models are used to explain biological pattern formation. Figures 2.2 and 2.3 depict aspects related to these processes. Figure 2.2 shows a schematic representation of a reaction-diffusion model, composed of an activator coupled to an inhibitor. If the diffusion strength is organized according to the pattern depicted in Figure 2.2(b), then this model undergoes phase transitions (symmetry breaking) during which it self-organizes at a more differentiated level. Figure 2.3 (taken from Meinhardt, 1982) presents the results of a simulation of a particular instance of a reaction-diffusion model. It shows how starting up the same model twice (panels b–d versus panels e–g) yields different networks. This is a type of variation (between panels d and g) which is not due to genetic or environmental influences but is solely due to the self-organizing growth of the system. Taken together, these figures demonstrate that heterogeneity in processes across time occurs in nature. In Molenaar, Boomsma, and Dolan (1993) this independent type of variation is called "third source" variation, alongside genetic and environmentally induced variation. It is conjectured

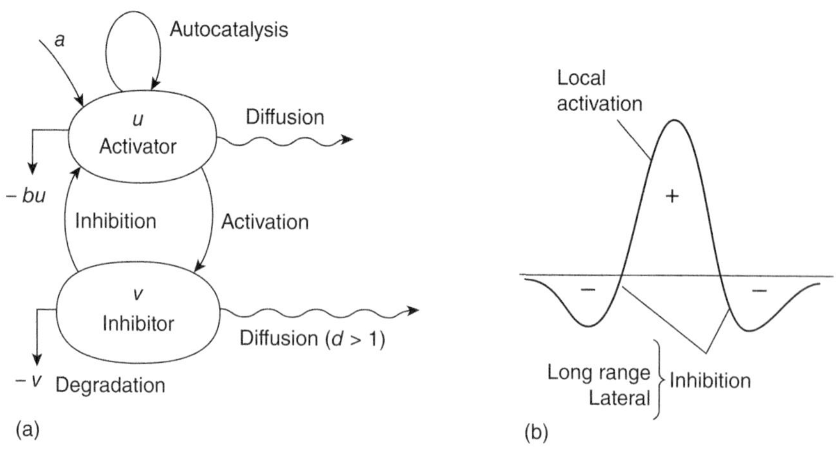

FIGURE 2.2
(a) Schematic representation of an activator-inhibitor reaction-diffusion system $u_t = a - bu + u^2/v + \nabla^2 u$, $v_t = u^2 - v + d\nabla^2 v$, where the subscript t denotes the partial derivative with respect to time and ∇^2 denotes the sum of second-order spatial derivatives with respect to the coordinates (Laplace operator). (b) Spatial representation of local activation diffusion ($\nabla^2 u$) and long-range inhibition diffusion ($d\nabla^2 v$); so-called on-center-off-surround pattern.

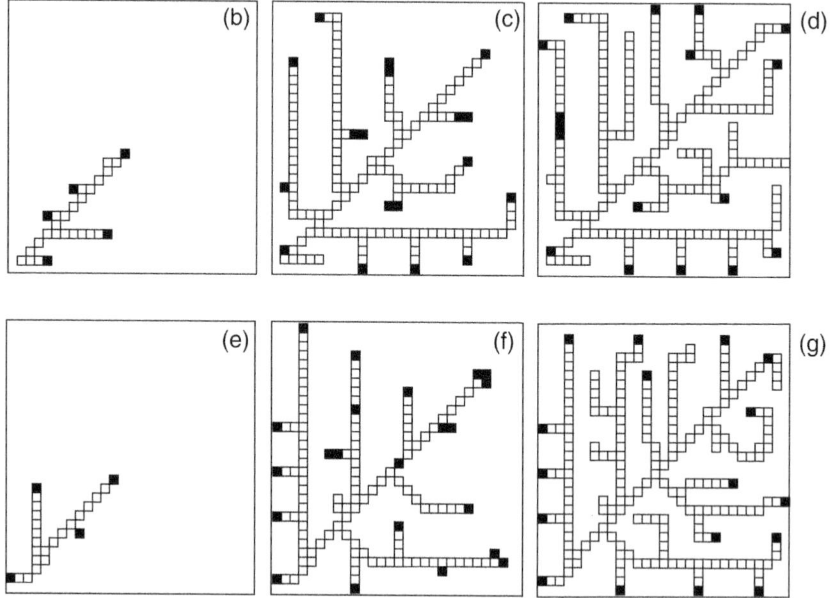

FIGURE 2.3
(b–d) and (e–g) Two simulations of an identical reaction-diffusion model (after fig. 15.5 of Meinhardt, 1982). Panels (b–d) show the development of a network in one simulation run; panels (e–g) show the development of a network in another simulation run using the same reaction-diffusion model.

that this third source variation gives rise to heterogeneous neural networks, which is the main cause for violations of criterion (b) for ergodicity (see Molenaar, 2007, for further details).

2.8 Example of a Non-Ergodic Process

For a univariate exemplar, we select the variable "Reckless" from the Borkenau data set. The reader can replicate the analysis using the online code (available through www.routledge.com/9781482230598). This item on the questionnaire asked participants on a scale of 0–6 how reckless they felt that day. We selected this variable since the average varied greatly across individuals. Hence, it already displays non-ergodicity. The average obtained when aggregating across all data (1.70, *sd* = 1.56) differs from the averages that are obtained for each individual separately. Figure 2.4 depicts the distribution of this variable for each of the 22 individuals. A noteworthy aspect of this data is that variability across time differs for each individual, with an average standard deviation of 1.06 (min = 0.45, max = 2.25). While a small majority (54%) has the aggregate average within the limits of their boxplots, a sizable proportion does not.

We motivated the discussion of ergodicity in terms of the mean for simplicity in explanation. As most researchers will be interested in the autocovariance function of a variable (and later, how multiple variable relate across time), we now present an example of heterogeneity in the lagged effects for individuals. A process can be said to be *autocovariance-ergodic* if the autocovariance estimate for an individual time series converges to the average estimate across time series as $T \rightarrow \infty$ (Porat, 2008). Differences across individuals emerged when examining the estimates of the autocovariance function at a lag of one (ACF(1)). A boxplot of the ACF(1) estimate for each individual is provided in Figure 2.5. Here, it appears there is some variability in the ACF(1) estimates with a range of –0.42 to 1.24. The interpretation of ACF estimates follows interpretation of standard covariance: for a negative (positive) estimate, when lag-1 values are higher than the mean the lag-0 (contemporaneous) variables are expected to be lower (higher for positive relations). The mean ACF(1) when calculated separately for each individual was 0.15, which would lead to inaccurate inferences if this was applied to all individuals—particularly those for whom there is a negative effect or a very large positive one. This example shows that the inference obtained when aggregating the data does not apply to the majority of individuals in this case.

One might argue that the process failed to converge to the ensemble average (i.e., average ACF(1) across all persons) because the sample size of $T = 90$ is relatively small and ergodicity is discussed in terms of an infinite number of

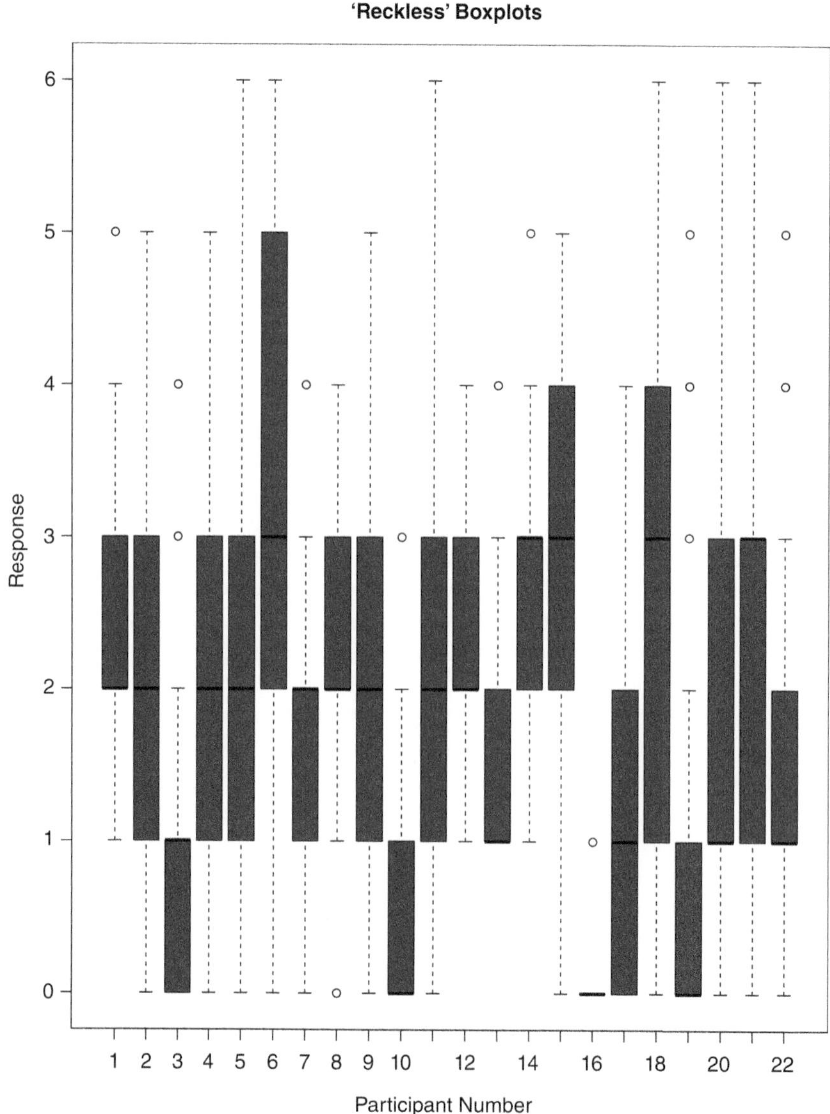

FIGURE 2.4
Distributions of values across individuals across time on the item "Reckless" from the Borkenau-Ostendorf data.

observations. Let us return to the example of the average to explore this. The individuals differed in their average across time on the Reckless variable. For the process to be ergodic, we must assume that given enough time all individuals will converge to the average level of Reckless. More generally, if given enough time all humans will converge to have the same psychological

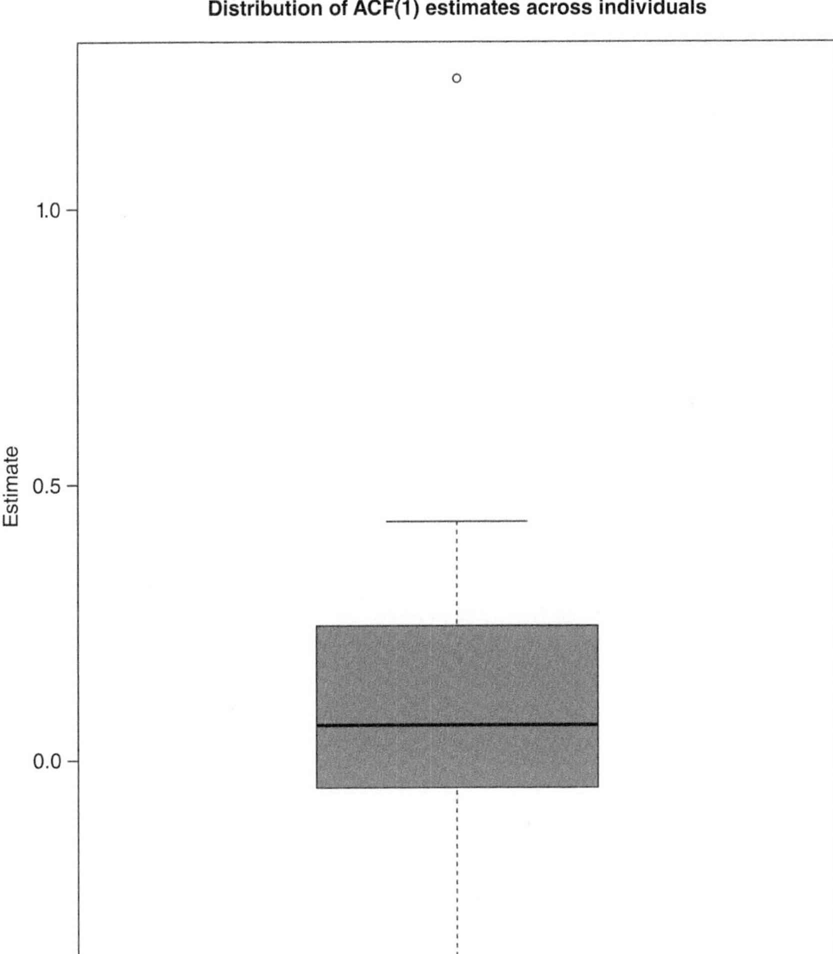

FIGURE 2.5
Distribution of individual-level ACF(1) estimates for each individual on the "Reckless" variable across the Borkenau-Orstendorf data.

processes. From what we know about human development, this proposition would require identical starting points and contexts. If we can learn from the biological systems in Figures 2.2 and 2.3, it is possible that human emotions and behaviors (such as Recklessness) which are formed by predispositions as well as context, are sensitive to starting points. Hence, an individual's average recklessness may differ from others with no sign of converging to the IEV mean across time. In fact, it is examination of such variation from the "norm" that is the foundation of the vast majority of psychological studies and research questions. In IEV analysis using correlation or regression, researchers examine precisely this variation from the norm (i.e., mean) and how that relates to variation in other variables.

What has been shown both numerically here and in prior work is that in finite cases individuals do not always exhibit the same patterns of behaviors, physiology, or emotions. Importantly, this follows what is seen in clinical practice as well as research studies on humans. For instance, clinical psychologists must attend to individual-level processes that they typically learn across time while a patient is in treatment (Fisher & Boswell, 2016). Another example is the intelligence quotient (IQ). This is known to differ across individuals, being sensitive to both genes (i.e., differing starting points; Plomin & von Stumm, 2018) and environment (i.e., context; Sauce & Matzel, 2018). Individuals may vary in IQ across their lifespan but if they are far from the average in either direction early in life, such as being intellectually disabled, it is not expected they will converge to the population mean if time were taken to infinity. Perhaps even more concretely, neuroimaging work has shown that individuals vary in the qualities of brain structures (e.g., Makris et al., 2015), physical connections among these brain structures (e.g., Barnea-Goraly et al., 2004), and functional aspects of brain processes (e.g., Finn et al., 2015; Laumann et al., 2015). At present, there is no work that suggests brain structures change and physical connections reorganize to some convergent mean for all humans if given enough time. In these ways the ergodic hypothesis is not typically supported for the study of human processes and definitively needs to be evaluated rather than taken as an assumption in analysis.

2.9 Conclusion

This chapter introduced foundational definitions of stationarity and related time series concepts necessary for understanding ergodicity. The classic ergodic theory covers the relation between results obtained in analysis of IEV and IAV. It was shown that for Gaussian processes, two necessary and sufficient criteria have to be met in order that such a relation exists (these two criteria are necessary, but not sufficient for general processes). The first criterion is the stationarity of the process, implying that no mean trends, time-varying variances, and/or lagged covariances must be present. The second criterion is homogeneity of the population, namely, the requirement that all individuals obey the same dynamical model. It was shown that standard analysis of IEV appears to be largely oblivious to the existence of widespread violation of this homogeneity assumption. That is, what is found cross-sectionally may not apply to individuals across time. A dedicated analysis of IAV is required to detect such violations.

References

Bakan, D. (1954). A generalization of Sidman's results on group and individual functions, and a criterion. *Psychological Bulletin, 51*(1), 63–64.

Barnea-Goraly, N., Kwon, H., Menon, V., Eliez, S., Lotspeich, L., & Reiss, A.L. (2004). White matter structure in autism: Preliminary evidence from diffusion tensor imaging. *Biological Psychiatry, 55*(3), 323–326.

Birkhoff, G. (1931). Proof of the ergodic theorem. *Proceedings of the National Academy of Sciences, USA, 17*(12), 656–660.

Boltzmann, L. (1885). Ueber die Eigenschaften monocyklischer und anderer damit verwandter Systeme. *Journal Fur Die Reine Und Angewandte Mathematik, 98*(1885), 68–94.

Box, G.E.P. & Jenkins, G.M. (1970). *Time series analysis: Forecasting and control.* San Francisco, CA: Holden-Day.

Brillinger, D. (1975). *Time series data analysis and theory.* New York, NY: Holt Rinehart.

Estes, W.K. (1956). The problem of inference from curves based on group data. *Psychological Bulletin, 53*(2), 134–140.

Fair, D.A., Bathula, D., Nikolas, M.A., & Nigg, J.T. (2012). Distinct neuropsychological subgroups in typically developing youth inform heterogeneity in children with ADHD. *Proceedings of the National Academy of Science, 109*, 6769–6774.

Finn, E.S., Shen, X., Scheinost, D., Rosenberg, M.D., Huang, J., Chun, M.M., & Constable, R.T. (2015). Functional connectome fingerprinting: Identifying individuals using patterns of brain connectivity. *Nature Neuroscience, 18*(11), 1664–1671. doi: 10.1038/nn.4135

Fisher, A.J. & Boswell, J.F. (2016). Enhancing the personalization of psychotherapy with dynamic assessment and modeling. *Assessment, 23*(4), 496–506.

Gottlieb, G. (1997). *Synthesizing nature-nurture: Prenatal roots of instinctive behavior.* Mahwah: Erlbaum.

Gottlieb, G. (1998). Normally occurring environmental and behavioral influences on gene activity: From central dogma to probabilistic epigenesis. *Psychological Review, 105*, 792–802.

Gottesman, I.I. & Gould, T.D. (2003). The endophenotype concept in psychiatry: Etymology and strategic intentions. *American Journal of Psychiatry, 160*(4), 636–645.

Hannan, E.J. (1970). *Multiple time series.* New York, NY: Wiley.

Hannan, M.T. (1991). *Aggregation and disaggregation in the social sciences.* Lexington, MA: Lexington Books.

Hayes, K.J. (1953). The backward curve: A method for the study of learning. *Psychological Review, 60*(4), 269–275.

Jenkins, G.M. & Watts, D.G. (1968). *Spectral analysis and its applications.* San Francisco: Holden-Day.

Kelderman, H. & Molenaar, P.C.M. (2007). The effect of individual differences in factor loadings on the standard factor model. *Multivariate Behavioral Research, 42*(3), 435–456.

Kolmogorov, A.N. (1933). *Foundations of probability.* New York, NY: University of Oregon.

Lamiell, J.T. (1987). *The psychology of personality: An epistemological inquiry.* New York, NY: Columbia University Press.

Lamiell, J.T. (2003). *Beyond individual and group differences: Human individuality, scientific psychology, and William Stern's critical personalism*. Thousand Oaks, CA: Sage.

Laumann, T.O., Gordon, E.M., Adeyemo, B., Snyder, A.Z., Joo, S.J., Chen, M.Y., ... & Petersen, S.E. (2015). Functional system and areal organization of a highly sampled individual human brain. *Neuron, 87*(3), 657–670.

Lerner, R.M. & Schmid, C.K. (2013). Relational developmental systems theories and the ecological validity of experimental designs. *Human Development, 56*, 372–380. doi: 10.1159/000357179.

Lindgren, G. (2013). *Stationary stochastic processes: Theory and applications*. Boca Raton, FL: CRC Press.

Lord, F.M. & Novick, M.R. (1968). *Statistical theories of mental test scores*. Reading, MA: Addison-Wesley.

Makris, N., Liang, L., Biederman, J., Valera, E.M., Brown, A.B., Petty, C., & Seidman, L.J. (2015). Toward defining the neural substrates of ADHD: A controlled structural MRI study in medication-naive adults. *Journal of Attention Disorders, 19*(11), 944–953.

Meinhardt, H. (1982). *Models of biological pattern formation*. London: Academic Press.

Merrill, M. (1931). The relationship of individual growth to average growth. *Human Biology, 3*, 37–70.

Molenaar, P.C.M., Boomsma, D.I., & Dolan, C.V. (1993). A third source of developmental differences. *Behavior Genetics, 23*(6), 519–524.

Molenaar, P.C.M. (2004). A manifesto on psychology as idiographic science: Bringing the person back into scientific psychology, this time forever. *Measurement: Interdisciplinary Research and Perspectives, 2*(4), 201–218.

Molenaar, P.C.M. (2007). On the implications of the classical ergodic theorems: Analysis of developmental processes has to focus on intra-individual variation. *Developmental Psychobiology, 50*(1), 60–69.

Molenaar, P.C.M. (2008). Consequences of the ergodic theorems for classical test theory, factor analysis, and the analysis of developmental processes. In S.M. Hofer & D.F. Alwin (Eds.), *Handbook of cognitive aging* (pp. 90–104). Thousand Oaks: Sage.

Molenaar, P.C.M. & Nesselroade, J.R. (2012). Merging the idiographic filter with dynamic factor analysis to model process. *Applied Developmental Science, 16*(4), 210–219. doi:10.1080/10888691.2012.

Nesselroade, J.R. & Ford, D.H. (1985). P-technique comes of age: Multivariate, replicated, single-subject designs for research on older adults. *Research on Aging, 7*(1), 46–80.

Nesselroade, J.R., Gerstorf, D., Hardy, S.A., & Ram, N. (2007). Idiographic filters for psychological constructs. *Measurement: Interdisciplinary Research and Perspectives, 5*(4), 217–235.

Plomin, R. & von Stumm, S. (2018). The new genetics of intelligence. *Nature Reviews Genetics, 19*(3), 148.

Porat, B. (2008). *Digital processing of random signals: Theory and methods*. Courier Dover Publications, Mineola, NY.

Porges, S.W. (2007). A phylogenetic journey through the vague and ambiguous Xth cranial nerve: A commentary on contemporary heart rate variability research. *Biological Psychology, 74*(2), 301–307.

Sauce, B. & Matzel, L.D. (2018). The paradox of intelligence: Heritability and malleability coexist in hidden gene-environment interplay. *Psychological Bulletin, 144*(1), 26.

Sidman, M. (1960). *Tactics of scientific research: Evaluating experimental data in psychology*. New York, NY: Basic Books.

Sidman, M. (1952). A note on functional relations obtained from group data. *Psychological Bulletin, 49*(3), 263–269.

Shumway, R.H. & Stoffer, D.S. (2000). *Time series analysis and its applications*. New York, NY: Springer-Verlag.

Smith, S.M. (2012). The future of fMRI connectivity. *NeuroImage, 62*(2), 1257–1266.

Volkmar, F.R., Lord, C., Bailey, A., Schultz, R.T., & Klin, A. (2004). Autism and pervasive mental disorders. *Journal of Child Psychology and Psychiatry, 45*(1), 135–170.

Wiener, N. (1930). Generalized harmonic analysis. *Acta Mathematica, 55*(1), 117–258.

Wright, A.G., Hopwood, C.J., & Simms, L.J. (2015). Daily interpersonal and affective dynamics in personality disorder. *Journal of Personality Disorders, 29*(4), 503–525.

3

P-Technique

In this chapter, we describe P-technique factor analysis of multivariate time series at the individual level. Factor analysis of any form seeks to estimate the relations between observed variables (or indicators, measures) and a latent construct that is thought to predict the observed variables. The estimated parameters can then be used directly for inference into how observed variables relate to latent constructs or to arrive at factor scores which represent variability in the latent construct of interest. For example, intelligence can be thought of as an unobservable attribute that causes test performance (Borsboom, Mellenbergh, & van Heerden, 2003). The observed indicators of the latent construct are considered to measure the latent construct of interest with error. Factor analysis directly models this measurement error (Bollen, 1989). Given that many constructs used in psychology are difficult to measure without error, such as depression or a child's inattentiveness, factor analysis has played a central role in the social sciences. Thus many readers may already be interested in factor analysis. P-technique is a person-specific factor analysis conducted across time for one individual.

Person-specific measurement is important for a few reasons. For one, perhaps some variables are better than others for indicating a specific construct for a given individual. Most diagnoses in the DSM-V require that only a proportion of possible symptoms be evident for a diagnosis to be made. This implies that not all individuals will exhibit all the symptoms. For instance, a diagnosis of Major Depressive Disorder (MDD) requires only five out of eight possible symptoms to be experienced for the past two weeks. If one wanted to assess depression levels for a given individual with an MDD diagnosis across time, one would want to measure that person as best as possible. Maybe for one person (person A), depressed feelings, weight gain, slowing down, fatigue, and feeling worthless captured their depression. A researcher or clinician would thus want to use these questions in the model. Person B may have their depression defined by diminished interest, weight losses, reduced activity, feeling guilty, and recurring thought of suicidal ideation. This person would require different questions from person A to assess the same construct.

A second reason for person-specific models is that the same observed item may relate differently to the latent construct for different people. Continuing with the above example, while both person A and B have "weight change" as an indicator of depression, for person A, it would have a positive relation with depression, whereas for person B, it would be negative to indicate

DOI: 10.1201/9780429172649-3

weight loss as an indicator of depression. Not only would each individual need to have different items used to assess their depression across time, but also the estimates for the way in which the items relate to the latent construct would have to be allowed to be person-specific in order to get the best estimate of depression at each time point for each individual.

The third reason for person-specific models is that it is possible that different constructs exist for different individuals. Perhaps for person A, questionnaire items pertaining to anxiety also relate to their overall depression scores, whereas for person B, anxiety items represent a distinct latent construct that is not a part of the depression latent construct. As these possibilities are well-known in the study of psychology, it is advised that measurement models be tailored to individuals. Methods such as P-technique which allow for exploratory analysis aids in the arrival of models that truly describe each individual.

P-technique was the first approach used to analyze intra-individual variation in psychology and aims to arrive at person-specific measurement models. It was introduced by Cattell (1943) and has since then been applied in numerous empirical studies, ranging from studies of affect (e.g., Garfein & Smyer, 1991; Lebo & Nesselroade, 1978; Mitteness & Nesselroade, 1987) to clinical studies (Cattell & Luborsky, 1950; Russell, Bryant, & Estrada, 1996) and other studies involving the analysis of other fluctuant constructs that varies in magnitude across states (e.g., Jones, Nesselroade, & Birkel, 1991; Roberts & Nesselroade, 1986; Schulenberg, Vondracek, & Nesselroade, 1988). Excellent reviews of applications of P-technique factor analysis can be found in Jones and Nesselroade (1990) and Russell, Jones, and Miller (2007). In this chapter, we present an effective approach grounded in structural equation modeling (SEM) for exploratory P-technique of single-subject multivariate time series.

3.1 The P-Technique Factor Model

Let $\mathbf{y}(t) = [y_1(t), y_2(t), \ldots, y_p(t)]'$ denote a p-variate time series for one individual observed at equidistant time points $t = 1, 2, \ldots, T$. The assumption of equidistance is required for most of the methods described in this book. The estimation of coefficients would be influenced by a violation of this assumption. Throughout this section, all time series are assumed to be weakly stationary. As discussed in Chapter 2, this implies that all time series have constant mean, variance, and (auto)covariance across time. In particular, the expected mean vector is constant across all points in time: $\mathrm{E}[\mathbf{y}(t)] = \mathrm{E}[\mathbf{y}(\mathrm{t}\text{-}u)] = \boldsymbol{\mu}_{\mathrm{y}} = [\mu_{y1}, \mu_{y2}, \ldots, \mu_{yp}]'$, where t-u denotes an arbitrary time point that differs from t. To ease the presentation, it again is assumed that all time series have zero mean function. In particular, $\mathrm{E}[\mathbf{y}(t)] = \boldsymbol{\mu}_{\mathrm{y}} = [0, 0, \ldots, 0]'$.

Consider the following P-technique factor model for $\mathbf{y}(t)$ introduced by Cattell (1943):

$$\mathbf{y}(t) = \mathbf{\Lambda}(0)\boldsymbol{\eta}(t) + \boldsymbol{\varepsilon}(t), \forall t. \tag{3.1}$$

In Equation (3.1), $\boldsymbol{\eta}(t)$ is a q-variate set of latent factor scores: $\boldsymbol{\eta}(t) = [\eta_1(t), \eta_2(t), \ldots, \eta_q(t)]'$ where $q < p$, and $\boldsymbol{\varepsilon}(t)$ is a p-variate measurement error process: $\boldsymbol{\varepsilon}(t) = [\varepsilon_1(t), \varepsilon_2(t), \ldots, \varepsilon_p(t)]'$ for the p-variate observed variables in $\mathbf{y}(t)$. The (p,q)-dimensional matrix $\mathbf{\Lambda}(0)$ is the matrix of factor loadings, or estimates for the prediction of the observed indicators from the latent factor. The factor loadings are constant for all time points. The $\boldsymbol{\eta}(t)$ factor scores do vary across time and represent the value for the latent construct (factor) at each time point. In words, the observed $y(t)$ variables equal the true scores ($\boldsymbol{\eta}(t)$) plus error ($\boldsymbol{\varepsilon}(t)$), as described in Lord and Novick (1968).

It is instructive to compare the P-technique factor model given by (3.1) with the standard factor model for the analysis of inter-individual variation. Let $\mathbf{y}_i$ denote the p-variate vector of observations obtained for a given subject $i = 1, 2, \ldots, N$: $\mathbf{y}_i = [y_{i1}, y_{i2}, \ldots, y_{ip}]'$. To ease the presentation, it is again assumed that the mean of $\mathbf{y}_i$ is the p-variate zero vector. Then, the standard factor model for y_i is

$$\mathbf{y}_i = \mathbf{\Lambda}\boldsymbol{\eta}_i + \boldsymbol{\varepsilon}_i, \forall i. \tag{3.2}$$

In (3.2), $\boldsymbol{\eta}_i$ is a q-dimensional vector of factor scores of subject i: $\boldsymbol{\eta}_i = [\eta_{i1}, \eta_{i2}, \ldots, \eta_{iq}]'$, and $\boldsymbol{\varepsilon}_i$ is a p-dimensional vector of measurement errors of subject i: $\boldsymbol{\varepsilon}_i = [\varepsilon_{i1}, \varepsilon_{i2}, \ldots, \varepsilon_{ip}]'$. The (p,q)-dimensional matrix $\mathbf{\Lambda}$ is the matrix of factor loadings, which are constant across individuals. Here, the variation in scores for the $\boldsymbol{\eta}_i$ is across people rather than across time. A comparison of (3.1) and (3.2) leads to the correspondences given in Table 3.1.

It is important to highlight that all random variables in the standard factor model (3.2), i.e., $\mathbf{y}_i$, $\boldsymbol{\eta}_i$, and $\boldsymbol{\varepsilon}_i$, are random variables across subjects $i = 1, 2, \ldots N$, considered at some implicit fixed time point. In contrast,

TABLE 3.1

Correspondences between the Standard Factor Model for Inter-Individual Variation and the P-Technique Factor Model for Intra-Individual Variation

	Standard Factor Model	P-Technique Factor Model
Observations	$\mathbf{y}_i$	$\mathbf{y}(t)$
Factor scores	$\boldsymbol{\eta}_i$	$\boldsymbol{\eta}(t)$
Measurement errors	$\boldsymbol{\varepsilon}_i$	$\boldsymbol{\varepsilon}(t)$
Loadings	$\mathbf{\Lambda}$	$\mathbf{\Lambda}(0)$

all random variables in the P-technique factor model, i.e., $\mathbf{y}(t)$, $\boldsymbol{\eta}(t)$, and $\boldsymbol{\varepsilon}(t)$, are random variables across time points, $t = 1, 2, \ldots, T$, pertaining to a fixed subject. This implies that the results obtained with the standard factor model (3.2) generalize to the population of subjects from which a random sample of N subjects has been drawn. In contrast, the domain of generalization for results obtained with the P-technique factor analysis of an observed time series $\mathbf{y}(t)$, $t = 1, 2, \ldots, T$ is the entire time axis (retrodiction to times $t < 1$, interpolation at $t = 1, 2, \ldots, T$, and prediction to times $t > T$) for a given i individual.

3.2 The Structural Model of the Covariance Function of $y(t)$ in P-Technique Factor Analysis

The structural model in P-technique factor analysis differs from that of standard factor analysis in important ways. We describe traditional factor analysis to provide the framework for comparison. Under the standard factor model (3.2) the estimated (p,p)-dimensional population covariance matrix $\boldsymbol{\Sigma}_y$ of the sample-level covariance matrix for $\mathbf{y}_i$ ($\mathbf{C}_y$) is obtained by pooling across *subjects*. In contrast, under the P-technique factor model (3.1) the estimated covariance matrix $\boldsymbol{\Sigma}_y(0)$ of the (p,p)-dimensional covariance matrix of $\mathbf{y}(t)$ at lag zero is obtained by pooling across *time* (see Table 3.2). As explained in Chapter 2, pooling across subjects is the hallmark of analysis of IEV, whereas pooling across time points is the hallmark of analysis of IAV. In either case, the covariance matrix of observed variables is used to obtain estimates.

The measurement model for the IEV (p,p)-dimensional population covariance matrix $\boldsymbol{\Sigma}_y$ in standard factor analysis is derived from Equation (3.2), using the assumption that the factor scores $\boldsymbol{\eta}_i$ and the measurement errors ε_i

TABLE 3.2

Correspondences between the Structural Covariance Model Underlying Standard Factor Analysis of Inter-Individual Variation and the Structural Covariance Model Underlying P-Technique Factor Model for Intra-Individual Variation

	Standard Factor Model	P-Technique Factor Model
Population covariance Matrix of y	Σ_y	$\Sigma_y(0)$
Sample covariance matrix of y	C_y	$C_y(0)$
Factor covariance	Ψ	$\Psi(0)$
Measurement error covariance	Θ	$\Theta(0)$

are uncorrelated: $\text{cov}[\boldsymbol{\eta}_i, \boldsymbol{\varepsilon}_i'] = \mathbf{0}$, where $\mathbf{0}$ is the (q,p)-dimensional zero matrix. It is then found that the measurement model for $\boldsymbol{\Sigma}_y$ is

$$\Sigma_y = \boldsymbol{\Lambda\Psi\Lambda}^T + \boldsymbol{\Theta}, \tag{3.3}$$

where $\boldsymbol{\Psi}$ is the (q,q)-dimensional covariance matrix of the factor scores: $\boldsymbol{\Psi} = \text{cov}[\boldsymbol{\eta}_i, \boldsymbol{\eta}_i']$. $\boldsymbol{\Theta}$ is the (p,p)-dimensional diagonal covariance matrix of the measurement errors: $\boldsymbol{\Theta} = \text{cov}[\boldsymbol{\varepsilon}_i, \boldsymbol{\varepsilon}_i']$. $\boldsymbol{\Theta}$ is diagonal because the standard definition for measurement errors in psychometrics is that they are conditionally independent: given the factor scores $\boldsymbol{\eta}_i$, measurement errors for two given observed variables y_k and y_j are uncorrelated across variables for $k \neq j$ conditional on the latent variable(s).

In P-technique factor analysis, it is similarly assumed that the latent factor series $\eta(t)$ and the measurement error process $\varepsilon(t)$ are independent at lag 0: $\text{cov}[\boldsymbol{\eta}(t), \boldsymbol{\varepsilon}(t)'] = \mathbf{0}$, where $\mathbf{0}$ is the (q,p)-dimensional zero matrix. With this assumption the structural model for the covariance function at lag zero, $\boldsymbol{\Sigma}_y(0)$, is derived from (1):

$$\Sigma_y(0) = \boldsymbol{\Lambda}(0)\boldsymbol{\Psi}(0)\boldsymbol{\Lambda}'(0) + \boldsymbol{\Theta}(0) \tag{3.4}$$

where the (q,q)-dimensional covariance matrix $\boldsymbol{\Psi}(0)$ is the covariance function of $\boldsymbol{\eta}(t)$ at lag zero and the (p,p)-dimensional covariance matrix $\boldsymbol{\Theta}(0)$ is the diagonal covariance function of $\boldsymbol{\varepsilon}(t)$ at lag 0. $\boldsymbol{\Theta}(0)$ is diagonal for the same reason as $\boldsymbol{\Theta}$ in Equation (3.3), namely, conditional independence. If we compare the covariance structure model (3.3) underlying standard factor analysis of inter-individual variation with the covariance structure model (3.4) underlying P-technique factor analysis, then it is evident that the two structural models are equivalent (see Table 3.2).

P-technique factor analysis is thus the factor analysis of time series and the same approach applied to cross-sectional data can be applied immediately to data that occurs across time within an individual. There is one major caveat in this approach. A p-variate time series $\mathbf{y}(t)$ is not characterized by a single (p,p)-dimensional covariance matrix like $\boldsymbol{\Sigma}_y$ in (3.3), but instead by a covariance *function* $\boldsymbol{\Sigma}_y(u)$, $u = 0, \pm 1, \ldots$ where u are integers denoting lags up to an arbitrarily larger order. The covariance function $\boldsymbol{\Sigma}_y(u)$ consists of a sequence of (p,p)-dimensional covariance matrices, one at each lag u. This means that the $\mathbf{y}(t)$ vectors of data across time are not independent, whereas the $\mathbf{y}_i$ vectors of data across individuals are independent across time. Consequently, factor analysis of $\mathbf{y}(t)$ *should* be based on a structural model for the entire covariance function $\boldsymbol{\Sigma}_y(u)$, $u = 0, \pm 1, \ldots$. However, this is not the way in which P-technique factor analysis is defined.

It is apparent from Equation (3.4) that P-technique factor analysis only specifies a structural model for part of the covariance function of a p-variate time series $y(t)$. Instead of yielding a structural model for the entire covariance function $\mathbf{\Sigma}_y(u)$, $u = 0, \pm 1, \ldots$, P-technique factor analysis only specifies a structural model for $\mathbf{\Sigma}_y(0)$, a (p,p)-dimensional covariance matrix associated with contemporaneous relationships (lag zero) among the p component series of $\mathbf{y}(t)$. Table 3.2 is thus obviously a severe simplification and, in this sense, P-technique factor analysis is an imperfect method; however, it provides a foundation for approaches that follow. In Chapter 5, we discuss approaches that accommodate the lagged components of the covariance function. The implications of this incompleteness of P-technique factor analysis will be addressed at the close of this chapter.

3.3 Conducting P-Technique Factor Analysis

The equivalence between the structure models associated with the standard factor model and the P-technique factor model implies that P-technique factor analysis can be carried out by means of standard factor analysis software. Throughout this section, we describe details necessary for carrying out exploratory P-technique analysis with SEM software. Integrated into this section are two illustrative applications of P-technique factor analysis—an application to simulated data and another application to empirical Borkenau-Ostendorf data obtained in an intensive repeated measurement design. We briefly describe the simulated data since it will be used to motivate the following descriptions of constraints necessary for estimating P-technique solutions. The online supplement contains code for the simulation analysis to follow (available via https://www.routledge.com/9781482230598).

3.3.1 Simulated Data

A zero mean six-variate ($p = 6$) time series $\mathbf{y}(t)$, $t = 1, 2, \ldots, T$ was simulated for $T = 200$ time points according to a P-technique two-factor model ($q = 2$). The (6,2) matrix of factor loadings $\mathbf{\Lambda}(0)$ and the (2,2)-dimensional factor covariance matrix $\mathbf{\Psi}(0)$ are given by (3.3):

$$\mathbf{\Lambda}(0) = \begin{bmatrix} 1 & 0 \\ 2 & 0 \\ 1 & 0 \\ 0 & 1 \\ 0 & 2 \\ 0 & 1 \end{bmatrix} \qquad \mathbf{\Psi}(0) = \begin{bmatrix} 2.77 & \\ 2.47 & 8.04 \end{bmatrix} \tag{3.5}$$

The (6,6)-dimensional measurement error covariance matrix $\mathbf{\Theta}(0)$ is a diagonal matrix: $\mathbf{\Theta}(0)$ = diag[0.5, 0.5, 0.5, 0.5, 0.5, 0.5]. Hence, the univariate

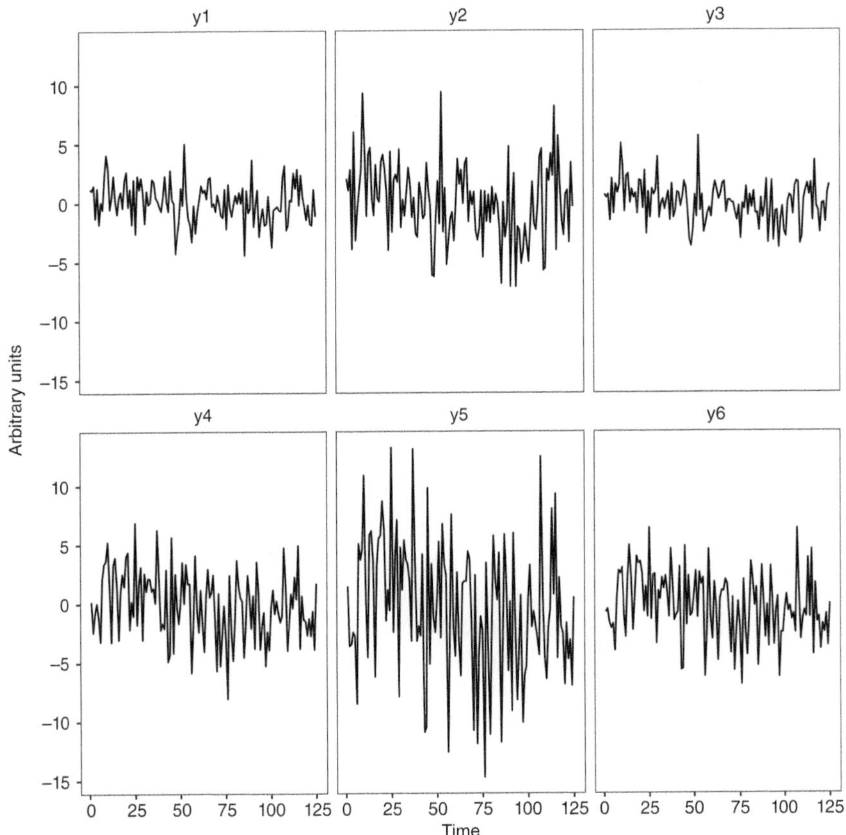

FIGURE 3.1
Depiction of observed multivariate time series matrix **Y**, $T = 200$, $p = 6$, simulated according to the P-technique two-factor ($q = 2$) model specified by (3.5).

measurement error processes $\varepsilon_k(t)$ and $\varepsilon_j(t)$ associated with different observed series $y_k(t)$ and $y_j(t)$ are uncorrelated: $\text{cov}[\varepsilon_k(t), \varepsilon_j(t)] = 0, k \neq j$. Figure 3.1 presents plots of the six univariate component series of $\mathbf{y}(t)$.

3.3.2 Constraints for Exploratory P-Technique Factor Analysis

Confirmatory P-technique analysis, where the structure of $\boldsymbol{\Lambda}_0$ is known, can be estimated using standard SEM software. When the structure is not known or when the researcher would like to investigate potential alterations to the structure exploratory analysis can be conducted. In an exploratory P-technique factor analysis the aim is to identify an instance of Equation (3.1), and hence Equation (3.4), yielding an acceptable fit to the data under the condition that we have no *a priori* information about the pattern of factor loadings in the matrix of factor loadings $\boldsymbol{\Lambda}(0)$. Having no *a priori* information about the pattern of factor loadings implies that we lack knowledge (theoretical or derived from prior data analyses) about: (a) the dimension q of the factor series $\eta(t)$ in (Equation 3.1); and (b) the way in which each univariate component factor series $\eta_k(t)$, $k = 1, 2, \ldots, q$, relates to the manifest component series $\mathbf{y}_j(t)$, $j = 1, 2, \ldots, p$. In other words, we do not know the structure indicating which observed variables relate to which latent construct(s). Hence, exploratory P-technique factor analysis involves answering points (a) and (b), where answering point (b) enables the conceptual interpretation (sometimes called labeling) of each component factor series $\eta_k(t)$, $k = 1, 2, \ldots, q$.

The following general strategy can be used to determine the dimension q of the latent factor series $\eta(t)$ in the P-technique factor model (3.1) for the six-variate time series $\mathbf{y}(t)$ depicted in Figure 3.1. First, a one-factor model is fitted to the data and its goodness of fit, or how well the data are explained by the model, is assessed. The way in which goodness of fit is assessed is explained shortly. If the fit of the one-factor model is deemed unsatisfactory, then a two-factor model is fitted to the data and its goodness of fit is assessed. If the fit of this two-factor model is deemed unsatisfactory, then a three-factor model is fitted, etc. The upper limit Q of factors in this sequence of potential q-factor models, $q = 1, \ldots, Q$, is a function of the dimension p of the manifest series $y(t)$. We will explain how Q can be determined shortly.

Exploratory P-technique q-factor models for a p-variate manifest time series $\mathbf{y}(t)$ require some constraints that can be defined in terms of (3.4) as follows: (a) the kth column of the (p,q)-dimensional matrix of factor loadings Λ_0 has $(k\text{-}1)$ loadings fixed at zero[1]; (b) the factor covariance matrix $\Psi(0)$ is the (q,q)-dimensional identity matrix; and (c) the (p,p)-dimensional measurement error covariance matrix $\Theta(0)$ is a diagonal matrix. Because all off-diagonal elements of $\Psi(0)$ are zero, the univariate latent component series $\eta_k(t)$, $k = 1, 2, \ldots, q$ are uncorrelated; hence, the analysis is an "orthogonal" exploratory factor analysis. The constraint indicated in (c) simply adheres to the aforementioned assumption that measurement errors are uncorrelated; this can be relaxed in some cases. Hence, for the six-variate time series $y(t)$ depicted in Figure 3.1, if $q = 1$ then $\Lambda(0)$ is a (6,1)-dimensional matrix (column vector) with six factor loadings to be estimated (so-called free factor loadings) and $\Psi(0)$ is the scalar 1.0. If $q = 2$, then $\Lambda(0)$ is a (6,2)-dimensional matrix with six free factor loadings in the first column and five $(k\text{-}1)$ free factor loadings in the second column (one factor loading in the second column being fixed at zero—it does not matter which one). In addition, $\Psi(0)$ is the (2,2)-dimensional identity matrix for scaling purposes. In both cases (i.e., a latent variable of 1 or 2), the $\Theta(0)$ matrix of measurement error covariances remains the same dimension of (p,p).

Now suppose that an exploratory P-technique three-factor model is fitted to the six-variate time series $\mathbf{y}(t)$ depicted in Figure 3.1. Then, $\Lambda(0)$ is a

[1] The method described here differs from traditional EFA approaches. The approach provided here circumvents some issues noted by Bollen (1989), such as enabling the researcher to inform the search by constricting some factor loadings to zero and specific correlations among observed and latent variables can be freed for estimation at the researcher's discretion.

(6,3)-dimensional matrix with six free factor loadings in the first column, five free factor loadings in the second column, and 4 free factor loadings in the third column. Hence, the total number of free factor loadings in the (6,3)-dimensional matrix Λ_0 is 6 + 5 + 4 = 15 "free" (or uniquely estimated) factor loadings. Also, the (6,6)-dimensional measurement error covariance matrix $\mathbf{\Theta}(0)$ has six measurement error variances along the diagonal, which together constitute an additional six free parameters to be estimated. The factor covariance matrix $\mathbf{\Psi}(0)$ is the (3,3)-dimensional identity matrix containing no free parameters. Consequently, the total number of free parameters in the P-technique three-factor model for the six-variate time series $\mathbf{y}(t)$ is 15 (free factor loadings) + 6 (free measurement error variances) = 21 free parameters.

This exploratory P-technique three-factor model has to be fitted to the (6,6)-dimensional covariance matrix $\mathbf{\Sigma}_y(0)$ in (3.4). $\mathbf{\Sigma}_y(0)$ contains six variances along the diagonal and 15 unique covariances off-diagonal, taking into account the symmetry of $\mathbf{\Sigma}_y(0)$. The general formula for the total number of variances and unique covariances in a (p,p)-dimensional covariance matrix is $p(p+1)/2$. Hence, $\mathbf{\Sigma}_y(0)$ contains $6(6+1)/2=21$ variances and covariances to be explained by a P-technique three-factor model containing 21 free parameters. It is evident that the exploratory P-technique three-factor model will yield an exact fit to the (6,6)-dimensional covariance matrix $\mathbf{\Sigma}_y(0)$. Exactly fitting factor models having the same number of free parameters as there are variances and covariances to be explained are called *saturated* models. The goodness of fit of a saturated model cannot be determined.

The count of free parameters in a factor model provides an upper limit Q of factors to be tested in exploratory orthogonal P-technique factor models. If the number of free parameters in a q-factor model is strictly less than the number of variances and covariances in the (p,p)-dimensional covariance matrix $\mathbf{\Sigma}_y(0)$ in (3.4), the model can be estimated and the resulting fit indices can be compared to models with different numbers of factors. Conversely, if the q-dimensional factor model contains more free parameters than the number of variances and covariances in $\mathbf{\Sigma}_y(0)$, then $Q < q$ and the model is said to be *under-identified*. For the (6,6)-dimensional covariance matrix $\mathbf{\Sigma}_y(0)$ associated with the 6-dimensional time series $\mathbf{y}(t)$ in Figure 3.1 it follows that $Q = q = 2$ can be fit, and the comparison for model fit between $q = 1$ and 2 can be conducted (but not $q = 3$).

Note that this method differs from traditional EFA approaches. The approach provided here circumvents some issues noted by Bollen (1989). First, the approach enables the researcher to inform the search by constricting some factor loadings to zero. Second, specific correlations among observed and latent variables can be freed for estimation at the researcher's discretion prior to conducting the exploratory search for factor structures. Allowing for more flexibility in the search process provides the opportunity for input from the researcher and prior work.

3.3.3 Assessing Goodness of Fit

Assessing goodness of fit aids in both evaluating the final model as well as in choosing the number of factors that best describe the data. In a standard factor analysis of inter-individual variation, the likelihood ratio test is the most powerful test to assess the goodness of fit of a factor model. In terms of (3.3), the likelihood ratio is the ratio of the likelihood of the covariance matrix implied by the factor model ($\mathbf{\Sigma}_y$) against the likelihood of the observed covariance matrix ($\mathbf{C}_y$). Large likelihood ratios indicate poor model fit. Negative two times the natural logarithm of this likelihood ratio is asymptotically χ^2 distributed (cf. Lawley & Maxwell, 1963). The chi-square distribution has degrees of freedom equal to the difference between the number of unique pieces of information in the covariance matrix of observed variables (using the equation introduced above, which counts unique variances and covariances), and the number of free parameters in the factor model. Hence, the number of degrees of freedom decreases as more parameters are added to a model. Given these properties, one can test if the likelihood ratio is significantly different than zero, which would suggest that the model provides a poor fit to the data. As with all statistical tests, the chi-square test is sensitive to sample size in that for a large sample size even small values can be significantly significant and indicate a poor fit. Another problem is that the likelihood will always improve when additional parameters are estimated. As such alternative measures of model fit have been developed and are highly used.

3.3.4 Alternative Indices of Model Fit

Many alternative indices to the chi-square test have been proposed to assess the fit of structural equation models to observed covariance matrices. The fit indices typically provide penalties for either the sample size or the number of parameters estimated. Most available software automatically provides numerous alternative fit indices. We will not go into specifics of these alternative indices given that much literature exists on the topic, but only consider the use of a few as they pertain to the evaluation of P-technique factor models.

To assess the performance of these indices in standard factor analysis of inter-individual variation several simulation studies have been performed. Brown (2006) presents a helpful summary of these simulation studies. Four alternative fit indices, namely, Standardized Root Mean Square Residual (SRMR; Hu & Bentler, 1999), Comparative Fit Index (CFI; Bentler, 1990), root mean squared error of approximation (RMSEA; Stieger & Lind, 1980), and Tucker-Lewis Index (TLI; Bentler & Bonett, 1980; Tucker & Lewis, 1973), performed overall the best in the simulation studies reviewed in Brown (2006). The TLI and CFI both penalize for adding additional numbers of parameters, with the TLI being slightly more strict. They are highly correlated and, usually, only one is reported. RMSEA also penalizes for the addition of parameters. SRMR does not penalize and is an absolute measure of fit with zero indicating a perfect fit. The RMSEA and SRMR tend to be positively biased which indicates a poorer fit—particularly for small sample sizes or small degrees of freedom.

Concise descriptions of these alternative fit indices are described elsewhere (in addition to the above citation, see Hu & Bentler, 1998; Tanaka, 1993).

Their use in assessing the fit of a structural equation model is different from the use of likelihood ratio tests. This is because the sampling distribution of the likelihood ratio test is known: the natural logarithm of the likelihood ratio is asymptotically chi-square distributed. Hence, one can carry out significance tests to determine whether the obtained value of the natural logarithm of the likelihood ratio exceeds a nominal alpha value. If this value exceeds the nominal alpha value, then the structural equation model is rejected as a model for the observed covariance matrix. In contrast, the sampling distributions of most of the alternative indices, in particular the four indices considered here, are unknown. Therefore, associated with each alternative index are criterion values which suggest adequate and good model fit. Specifically, For SRMR, a value less than 0.08 suggests adequate model fit; a value less than 0.05 suggests a good model fit. Similarly, low values of RMSEA indicate a good fit, with 0.10 suggesting an adequate fit and 0.05 an excellent fit. For CFI, a value larger than 0.90 suggests adequate model fit; a value larger than 0.95 suggests a good model fit. Finally, for TLI, a value larger than 0.90 suggests an adequate model fit; a value larger than 0.95 suggests a good model fit (Brown, 2006).

3.3.5 An Important Caveat

The asymptotic χ^2 distribution of the likelihood ratio test for the standard factor model (3.2) described above is derived from Lawley and Maxwell (1963). With respect to the IEV factor model, it can be assumed that observed p-variate vectors $\mathbf{y}_i$ and $\mathbf{y}_k$ for different subjects $i \neq k$ are mutually independent: $\text{cov}[\mathbf{y}_i, \mathbf{y}_k'] = \mathbf{0}$. Thus the estimate of the (p,p)-dimensional covariance matrix $\mathbf{C}_y$ consists of a sum over N independent components $\mathbf{y}_i\mathbf{y}_i'$, $i = 1, \ldots, N$ (remember that $\mathbf{y}_i$ is zero mean). In contrast, the estimate of the (p,p)-dimensional covariance matrix $\mathbf{C}_y(0)$ consists of a sum over *T-dependent* components $\mathbf{y}(t)\mathbf{y}(t)'$, because $\mathbf{y}(t)$ is a (zero mean) weakly stationary time series for which $\text{cov}[\mathbf{y}(t_1), \mathbf{y}(t_2)'] \neq \mathbf{0}$ if $t_1 \neq t_2$ in many cases. Hence, it would seem to be uncertain whether the likelihood ratio test in P-technique factor analysis has the same asymptotic chi-square distribution as derived in Lawley and Maxwell (1963) given this lack of independence in the sequential observations.

In a similar line of thought, P-technique was met with severe criticism by the eminent statistician T.W. Anderson (1963). The basic tenet of this criticism is that the use of standard factor analytic techniques developed for the analysis of inter-individual variation (where replications are assumed to be measurement independent) cannot be validly applied to the analysis of sequentially dependent intra-individual variation (time series). This criticism by one of the most distinguished statisticians of that time has had a strong negative influence on the further development of P-technique. In this section, it will be argued that a wholesale disqualification of P-technique that appeared to be implied by Anderson's (1963) criticism is not warranted. That is, P-technique as originally conceived by Cattell (1951) can validly be applied for certain restricted purposes. In particular, P-technique is shown to yield reliable estimates of latent factor series, even if the observed series is highly sequentially dependent.

3.3.5.1 The Recoverability of P-Technique

Simulation studies suggest that the unmodeled lagged relations do not present a problem when arriving at the data-generating structure of the measurement model. First, Molenaar and Nesselroade (1998) presented results obtained in a small simulation study showing that two well-known estimation techniques (quasi-maximum likelihood estimation and asymptotically distribution-free estimation) both appear to yield consistent parameter estimates in P-technique factor analysis even when the data have serial dependencies. Asymptotically distribution-free estimation also appears to yield consistent likelihood ratio tests and standard errors of parameter estimates. For relatively short observed time series ($T \leq 100$) both estimation methods yield very similar likelihood ratio test values and parameter standard errors.

In a follow-up study, Molenaar and Nesselroade (2009) addressed the question: How and how much does P-technique factor analysis distort the character of one's data when it is applied to multivariate time series containing lagged relations? They generated highly sequentially dependent multivariate time series data by means of a state space model (to be introduced formally in the next chapters) and subjected these data to P-technique factor analysis. It was found that the factor loadings obtained with P-technique closely correspond with their true values in the simulation model. Moreover, importantly, the estimated latent factor series correlate almost perfectly with the true factor series used in the simulation. This latter finding provides the underpinning for an approximate dynamic factor analysis procedure that will be introduced in Chapter 5.

3.3.5.2 Statistical Theory

Wooldridge (1994) presents a comprehensive review of the asymptotic statistical theory of parameter estimation in time series models. In a nutshell, Wooldridge (1994) shows that for weakly stationary time series similar asymptotic results apply as for the independently identically distributed observations figuring in Lawley and Maxwell's (1963) derivation. This is an important result because in P-technique, it is assumed that $y(t)$ is weakly stationary.

3.3.5.3 Concluding Thoughts

Despite the positive asymptotic results reviewed in Wooldridge (1994) and the simulation results, we have to be careful and conservative in the interpretation of likelihood ratio tests of P-technique factor models. To emphasize this, these likelihood ratio tests will be referred to as *quasi-likelihood ratio tests* (referred to elsewhere as *pseudo-Maximum Likelihood*), in recognition of the fact that the data are, in general, sequentially dependent. In particular, we will put more emphasis on quasi-likelihood ratio tests comparing different P-technique models fitted to the same data (so-called χ^2 difference tests) because these can be expected to be unaffected by any constant bias. In contrast, quasi-likelihood ratio tests for evaluating the goodness of fit of any particular P-technique model compared to the observed sample covariance matrix (i.e., chi-square goodness of fit statistics) will be used as supplementary information for diagnostic purposes. Moreover, we will also consider alternative indices of model fit which may be less affected by the sequential dependency of the data.

3.3.6 Convention

To reiterate, likelihood ratio tests in P-technique factor analysis will be referred to as quasi-likelihood ratio tests. The natural logarithm of the quasi-likelihood ratio test will be denoted by *ln qLR*. The ln qLR is treated in exactly the same way as the natural logarithm of the likelihood ratio in standard factor analysis of inter-individual variation. That is, the asymptotic chi-square distribution of the natural logarithm of the likelihood ratio in standard factor analysis will also be assumed to hold for the ln qLR. In particular, significance tests in P-technique factor analysis based on the ln qLR will be carried out in the same way as significance tests based on the natural logarithm of the likelihood ratio in standard factor analysis.

3.3.7 Determining the Number of Factors in P-Technique Factor Analysis

Until now we have given much attention to fit indices. Now we bring this information together to aid in selecting the best model describing observed time series data. Lo, Molenaar, and Rovine (2017) carried out a large-scale simulation study to determine the performance of 10 different criteria to determine the dimension of the latent factor series in P-technique factor analysis. Sequentially dependent multivariate time series were generated by means of state space models (which will be explained in Chapter 5) under various conditions such as the varying dimension of the factor series, and length of the generated series. The criteria included alternative fit indices such as SRMR, TLI, RMSEA, and CFI. Additional indices based on eigenvalues of the estimate of $\mathbf{\Sigma}_y(0)$ in (3.4), such as the acceleration factor (AF) criterion, were also tested. The commonly used AF criterion approximates the second derivative evaluated at each eigenvalue in the plot of eigenvalues ordered according to decreasing rank (the so-called scree-plot) and places cutoff values based on this scree curvature (Raîche et al., 2013). It was found that the SRMR and the AF criteria perform best in terms of the proportion of correctly identified dimensions of factor series and bias in P-technique.

Another paper focusing on selecting the optimal number of factors in time series data arrived at different inferences (Lu, Chow, & Loken, 2017). The simulations tested performance in selecting cross-loadings under a number of conditions: sample size, cross-loading sizes, and violations of distributional assumptions. Here, the model fit indices previously discussed and Likelihood ratio tests were compared against Bayesian model selection criteria. They found that the Bayesian information criterion (BIC; Schwarz, 1978) and the related Bayes factor (BF; Goodman, 1999) appropriately recovered the correct dimensionality. These criteria compare the fit of one model to another model, and thus can be useful when trying to identify which model better explains the data in relative terms. Rather than offering an absolute measure, the BIC is scaled according to the data and thus no thresholds representing "good fit" exist. The BIC is a relative measure and the model with the lowest BIC is said to be the best. The paper also found that comparing the change in likelihood of two nested models and the RMSEA performed particularly well. Taken together, in practice, multiple criteria should be simultaneously considered when conducting exploratory P-technique analysis. We present examples here using both simulated and empirical data.

We demonstrate the selection of the number of latent variables with the simulated data. The fit of a P-technique one-factor model to the six-variate series depicted in Figure 3.1 yields the following results. The obtained ln qLR is 647.74, chi-square distributed with degrees of freedom = 9, $p < 0.0001$. Here, p can be interpreted as the probability that the data have been generated by the postulated P-technique one-factor model. Of the alternative fit indices, we obtained: SRMR = 0.26; TLI = 0.10; CFI = 0.46; RMSEA = 0.75; and BIC = 3231.84. The ln qLR has a probability smaller than any reasonable nominal alpha (e.g., $\alpha = 0.05$ or $\alpha = 0.01$), indicating a poor fit. The SRMR is larger than 0.08; the TLI is smaller than 0.90; the RMSEA is much larger than 0.10; and the CFI also is smaller than 0.90. Hence, the likelihood ratio test and all three alternative indices indicate that the hypothesis that the data have been generated according to a P-technique one-factor model has to be rejected.

The exploratory P-technique two-factor model has $q = 2$ orthogonal factors. That is, $\mathbf{\Psi}(0)$ in (3.4) is defined as the (2,2)-dimensional identity matrix. Also, the (6,2)-dimensional matrix of factor loadings $\mathbf{\Lambda}(0)$ has one element of the second column fixed at zero. It does not matter which element of the second column of $\mathbf{\Lambda}(0)$ is fixed at zero, e.g., $\lambda_{12} = 0$. The ln qLR is 1.26, degrees of freedom = 4, $p = 0.87$; SRMR = 0.00; RMSEA = 0.00; TLI = 1.00; CFI = 1.00; BIC = 2609.50. The ln qLR has a probability larger than any reasonable nominal alpha; the SRMR is smaller than 0.05; the TLI is larger than 0.95; the CFI also is larger than 0.95; and the BIC is smaller than the previous model. Hence, the quasi-likelihood ratio test and all alternative indices indicate that the hypothesis that the data have been generated according to a P-technique two-factor model cannot be rejected. Of note, in this case, the SRMR, parallel analysis, BIC, and AF suggested a two-factor solution (see online supplemental). Hence, by all standards reported from prior large-scale simulation studies, one can be confident in this conclusion. In practice, these measures may not all point to the same factor solution. Researchers must then look closer into the qualities of the data (e.g., number of observations, number of variables) and investigate which approach is best for the data qualities.

3.3.8 Oblique Rotation to Simple Structure

Having selected the two-factor model for the data depicted in Figure 3.1, the exploratory orthogonal P-technique factor analysis is complete. Remember that in the exploratory orthogonal P-technique factor analysis the (2,2)-dimensional covariance matrix $\mathbf{\Psi}(0)$ in (3.4) is defined as the identity matrix, which implies that the 2 latent variables in $\boldsymbol{\eta}(t) = [\eta_1(t), \eta_2(t)]'$ in (3.1) are uncorrelated: $\psi_{12}(0) = \text{cov}[\eta_1(t), \eta_2(t)] = 0$. These orthogonal factors often have a pattern in the $\hat{\boldsymbol{\Lambda}}(0)$ factor loading matrix whereby the observed univariate series $\mathbf{y}_k(t)$, $k = 1, 2, \ldots, 6$, for the variables in $\mathbf{Y}(t)$ may have substantial factor loadings λ_{k1} and λ_{k2} on both $\eta_1(t)$ and $\eta_2(t)$. This can make interpretation of the factors difficult.

Oftentimes researchers prefer what is called a *simple structure,* although some argue it is not optimal (Tucker & Lewis, 1973). Simple structure implies that each univariate observed component series $\mathbf{y}_k$ is a pure indicator of either $\eta_1(t)$ or $\eta_2(t)$ but not both. For example, if a given variable $\mathbf{y}_k$ has a large absolute value on λ_{k1} (i.e., loads on factor 1), the factor loading for λ_{k2} would equal zero (or its absolute value be close to zero) if a simple structure were present. Of course, in some cases, indicator variables may load onto more than one latent variable. Nonetheless, moving toward a simpler structure oftentimes aids in interpretation of the factors by enabling statements regarding which observed variables load onto which latent variable.

By allowing the latent factors to correlate a simple (or simpler) structure can sometimes be obtained. This process, referred to as, "direct oblimin rotation", is considered an *oblique rotation* and is contrasted against *orthogonal* models wherein the factors do not correlate. Other rotation methods are available, the two most common being *promax* and *varimax.* Promax rotation is another oblique method that is computationally faster than direct oblimin approaches. Thus, promax is helpful when the data sets are very large. Varimax rotation is an orthogonal option that sums the variances of the squared loadings. The goal is to arrive at as close to a simple structure as the data will allow.

The (6,2)-dimensional matrix of estimated factor loadings, $\hat{\boldsymbol{\Lambda}}(0)$, thus obtained via the exploratory orthogonal P-technique two-factor model on our example data does not have a simple structure:

$$\hat{\boldsymbol{\Lambda}}(0) = \begin{bmatrix} 1.58 & 0^* \\ 3.23 & -0.29 \\ 1.60 & -0.11 \\ 1.43 & 2.51 \\ 2.86 & 4.97 \\ 1.54 & 2.49 \end{bmatrix} \tag{3.6}$$

The superscript * denotes the loading was fixed at that value. $\hat{\boldsymbol{\Lambda}}(0)$ is now subjected to oblique factor rotation to a simple structure by running a semi-confirmatory model wherein the off-diagonal element(s) in $\boldsymbol{\Psi}(0)$ are freed for estimation. A loading on the first factor is set to 0 to ensure that each factor has at least one indicator that loads uniquely onto it. This can be chosen by selecting the lowest lambda value obtained in the orthogonal case, as is done here, or selecting an indicator based on prior research or hypotheses. The rotated matrix of factor loadings $\widehat{\boldsymbol{\Lambda}}^{R}(0)$ thus obtained, as well as the associated covariance matrix $\widehat{\boldsymbol{\Psi}}^{R}(0)$ of the obliquely rotated factor series $\eta^{R}(t)$:

$$\widehat{\boldsymbol{\Lambda}}^{R}(0)=\begin{bmatrix} 1.58 & 0^{*} \\ 3.40 & -0.34 \\ 1.66 & -0.12 \\ 0^{*} & 2.89 \\ 0.04 & 5.72 \\ 0.12 & 2.87 \end{bmatrix} \quad \widehat{\boldsymbol{\Psi}}^{R}(0)=\begin{bmatrix} 1^{*} & \\ 0.49 & 1^{*} \end{bmatrix} \tag{3.7}$$

The diagonal elements of $\widehat{\boldsymbol{\Psi}}^{R}(0)$ have superscript * to indicate that these values have been fixed at 1.0 to assign scales to the latent factor series. An alternative would be to fix the factor loading for the kth variable $\mathbf{y}_k(t)$ to be 1; this would then scale the latent variable to match that of the kth variable. It must be scaled at either the observed or latent variable levels to identify the metrics of the latent variables (Bollen, 1989). Scaling the latent variables to 1 enables the comparison of lambda values across the factors.

3.3.9 Testing the Final Oblique P-Technique Two-Factor Model

The obliquely rotated (6,2)-dimensional matrix $\widehat{\boldsymbol{\Lambda}}^{R}(0)$ has a pattern of factor loadings that correspond more closely to a simple structure than $\widehat{\boldsymbol{\Lambda}}$. In particular, it appears that the elements λ_{41}, λ_{51}, λ_{61}, λ_{12}, λ_{22}, and λ_{32}, are all close to zero. Fixing these four-factor loadings to zero constitutes a fully final pattern for the matrix of factor loadings $\widehat{\boldsymbol{\Lambda}}_c(0)$ in a confirmatory oblique P-technique two-factor model.

The final oblique P-technique two-factor model is fitted to the data depicted in Figure 3.1. This time, fixing one lambda value in each factor to one scales the latent variables. That is, the latent factor scores will now be scaled according to the scaling of this variable and reflect the data-generating model, which had variances in the factors that were greater than 1.

The ln qLR for the confirmatory model is 8.88 with 8 degrees of freedom, $p = 0.35$; SRMR = 0.02; RMSEA = 0.03; TLI = 1.00; CFI = 1.00. The quasi-likelihood ratio test has a probability larger than any reasonable nominal alpha; the SRMR and RSMEA are smaller than 0.05; the TLI is larger than 0.95; the CFI also is larger than 0.95. Hence, the likelihood ratio test and all alternative indices indicate that the hypothesis that the data have been generated according to this confirmatory oblique P-technique two-factor model cannot be rejected. The model appears to describe the data well.

But more can be said. The ln qLR of the exploratory orthogonal P-technique two-factor model was found to be 2.50 with degrees of freedom = 4. The ln qLR of the confirmatory oblique P-technique two-factor model is as follows: ln qLR = 8.57, degrees of freedom = 8. Hence, the ln qLR of the confirmatory oblique model with respect to the exploratory orthogonal model is Δln qLR = 8.57 − 2.50 = 6.07. This Δln qLR is considered to be asymptotically chi-square distributed with 8 − 4 = 4 degrees of freedom. A difference of 6.07 with 4 degrees of freedom has a probability of occurring that is higher than most alpha levels used in research. Hence, the fit of the oblique confirmatory model does not differ significantly from the fit of the orthogonal exploratory model. Because the confirmatory oblique model has four free parameters less than the exploratory orthogonal model, the confirmatory oblique model is the preferred model.

The estimated (6,2)-dimensional matrix of factor loadings $\widehat{\boldsymbol{\Lambda}}_c(0)$ and the estimated covariance matrix $\widehat{\boldsymbol{\Psi}}_c(0)$ are given below. The subscript *c* highlights that these estimated matrices were associated with the confirmatory oblique model. The entries in $\widehat{\boldsymbol{\Lambda}}_c(0)$ with superscript * have been fixed. The two factor loadings fixed at 1.0 assign scales to the latent factor series.

$$\widehat{\boldsymbol{\Lambda}}_c(0) = \begin{bmatrix} 1^* & 0 \\ 2.05 & 0 \\ 1.02 & 0 \\ 0 & 1^* \\ 0 & 1.99 \\ 0 & 1.01 \end{bmatrix} \qquad \widehat{\boldsymbol{\Psi}}_c(0) = \begin{bmatrix} 2.48 & \\ 2.01 & 8.34 \end{bmatrix} \tag{3.8}$$

A comparison of $\widehat{\boldsymbol{\Lambda}}_c(0)$ and $\widehat{\boldsymbol{\Psi}}_c(0)$ with their true values used to simulate the data given by (3.5) shows that the estimated factor loadings are close to their true values. The one discrepancy is that the estimated covariance of $\eta_2(t)$ and $\eta_1(t)$ at lag zero, $\psi_{c,21}(0) = 2.01$, is low in comparison with its true value of $\psi_{21}(0) = 2.47$. The path diagram of the confirmatory oblique P-technique two-factor model is given in Figure 3.2.

3.3.10 Empirical Example

Data gathered for research purposes oftentimes do not provide clear solutions. For an example of what researchers may encounter in practice, we use data from one of the individuals, participant number 5, in the Borkenau and Ostendorf (1998) data set. Code can be found in the online supplement. As described in Chapter 1, there are 30 indicators of the Big 5 personality inventory that were collected daily for 90 days.

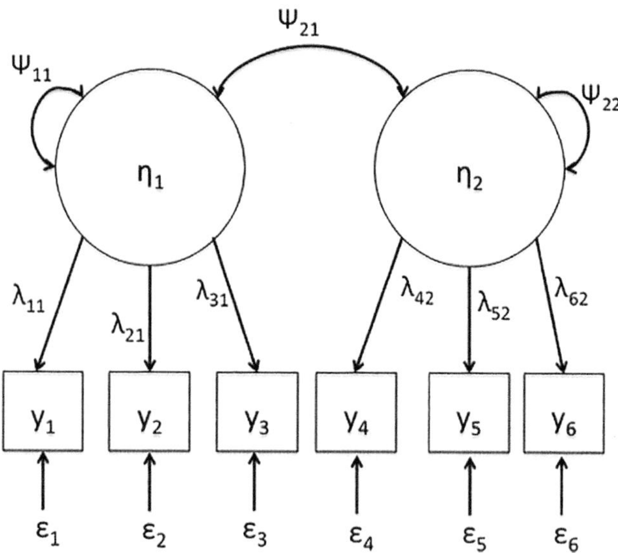

FIGURE 3.2
Depiction of an oblique two-factor model.

Analysis begins by conducting exploratory factor analysis as indicated above. Table 3.3 provides the fit index results for the orthogonal factor solution. We see that with empirical data, the number of factors is not always clear. For instance, all indices suggest an acceptable model fit at a three-factor solution. This is also the solution with the lowest BIC, which Lu et al. (2017) found to typically provide the data-generating solution in simulation studies. The SRMR values also are considered good for the three-factor solution, which aligns with the recommendation of Lo et al. (2017). However, we see improvement in model fits according to the CFI, TLI, RMSEA, and SRMR as the number of factors increases. Given that the BIC opts for the three-factor solution and fit indices are acceptable, it is a defendable decision to use the three-factor solution.

Following this, the process described for the simulated data example was repeated. That is, the orthogonal three-factor solution was inspected and the maximum loading for each factor was identified. The associated variable was

TABLE 3.3
Fit Indices from Exploratory Factor Analysis on Borkenau and Ostendorf (1990) Data

	Chi²	df	*p*-value	CFI	TLI	RMSEA	SRMR	BIC
1 Factor	1452.63	405	0	0.69	0.67	0.17	0.13	9159.90
2 Factors	790.78	376	0	0.88	0.86	0.11	0.06	8628.55
3 Factors	545.03	348	0	0.94	0.93	0.08	0.03	8508.79
4 Factors	478.86	321	0	0.95	0.94	0.07	0.03	8564.11
5 Factors	421.51	295	0	0.96	0.95	0.07	0.02	8623.76

not included in the other factors for the next step: estimate an oblique solution by allowing the factors to correlate. The latent variable was scaled to have a variance of one, enabling investigation into relative weights of factor loadings. Here, indicators with factor loadings with an absolute value greater than 0.30 were retained. The results can be found in Table 3.4.

TABLE 3.4

Loading Estimates from the Final Factor Solution for person 6 of the Borkenau-Orstendorf Data

	Attribute	FA1	FA2	FA3
V1	Lazy	0.84	--	--
V2	Dynamic	−0.69	0.41	--
V3	Selfish	--	0.64	0.76
V4	Unimaginitive	0.58	−0.55	--
V5	Irritable	--	−0.60	0.45
V6	Industrious	−0.95	--	--
V7	Emotionally stable	−0.45	--	−0.56
V8	Witty	−0.42	--	−0.37
V9	Calm	0.42	--	−0.61
V10	Good natured	--	--	−0.73
V11	Domineering	−0.60	0.76	0.64
V12	Helpful	−0.62	--	--
V13	Knowledgeable	−0.86	--	--
V14	Prudent	−0.77	--	--
V15	Persistent	−0.52	--	--
V16	Sociable	−0.31	0.77	--
V17	Bad-tempered	0.31	--	0.50
V18	Reckless	--	0.85	--
V19	Fanciless	0.46	−0.64	--
V20	Obstinate	−0.49	0.73	0.62
V21	Considerate	--	−0.60	−0.83
V22	Shy	0.39	−0.72	--
V23	Silent	0.36	−0.76	--
V24	Lively	--	0.79	--
V25	Changeable	0.90	--	--
V26	Resistant	--	0.57	--
V27	Uninformed	0.83	--	--
V28	Responsible	0.88	--	--
V29	Vulnerable	--	−0.82	--
V30	Reserved	0.33	−0.75	--

A few noteworthy aspects become apparent. For one, this is not a simple structure. Numerous indicators load onto more than one latent construct. This is expected given prior work which has found very different factor solutions across individuals using this data. It is likely a symptom of the personality features not being clear indicators of distinct latent constructs. Still, some meaning can be teased out from the solution. Factor two, containing high loadings on *Sociable, Dynamic, Reckless, Lively,* and *Domineering* and negative loadings on attributes such as *Silent*, paints a picture of a state where one is engaged with others—albeit perhaps in a negative manner. Factor one depicts a solitary state, with attributes such as *Lazy* and *Reserved* loading highly. Another aspect is that numerous loadings are negative. If a simple solution were provided, one might want to reverse-code such variables. However, given that variables tend to be negative for one factor and positive for another, the need for reverse-coding is obviated.

3.4 Conclusion

In this chapter, it was explained how a p-variate time series obtained with a single participant is to be subjected to exploratory P-technique factor analysis. That is, P-technique factor analysis under the condition that no *a priori* information is available regarding (a) the dimension q of the latent factor series and (b) the way in which the factor series affects the observed series. P-technique is useful when one wishes to obtain person-specific measurement models in an effort to better assess individuals' scores on latent constructs.

3.4.1 Statistical Background

We demonstrated here how the unknown dimension q of the latent factor series underlying the observed series is determined in a sequence of exploratory orthogonal P-technique q-factor models, starting with $q = 1$ and increasing q by one each time the current model does not yield a satisfactory fit. This stepwise procedure may either result in a satisfactorily fitting model with q-dimensional latent factor series, or else may not yield any satisfactorily fitting factor model in case q becomes larger than the upper limit Q, where Q is the largest dimension of the latent factor series that still yields an identifiable exploratory orthogonal P-technique factor model.

If a satisfactorily fitting exploratory orthogonal P-technique q-factor model is found, then it has to be determined how the factor series affects the observed series. The important heuristic tool to accomplish this is to obliquely rotate the (p, q)-dimensional matrix of factor loadings to a simple structure. A matrix of factor loadings has a pure simple structure if each observed univariate component series $\mathbf{y}_k(t)$, $k = 1, 2, \ldots, p$, only has a single nonzero factor loading on one of the q univariate component factor series $\eta_l(t)$, $l = 1, 2, \ldots, q$. Of course, a simple structure thus defined is an ideal that in analyses of empirical data can only be approached to some degree.

If successful, the oblique rotation to simple structure yields a final pattern for the (p,q)-dimensional rotated matrix of factor loadings that approaches the ideal of simple structure as closely as possible. This final pattern of factor loadings implies that the small factor loadings can be fixed at zero. The resulting oblique P-technique q-factor model then is fitted to the data in order to test its goodness of fit. If the fit of the final model is satisfactory, the P-technique factor analysis is completed. If, however, the fit of this model is not satisfactory then additional steps have to be carried out to remedy this. These additional steps are explained in the next section on confirmatory P-technique factor analysis.

3.4.2 Application of P-Technique to Empirical Data Sets

We conducted exploratory P-technique to the Borkenau and Orstendorf ecological momentary assessment data example. Here, the solution was not as clean as our simulated example and the factors did not correspond with the latent constructs used to derive the measure (i.e., personality traits). As seen here, it is possible that some sets of indices will not provide clear measurement solutions at the individual level. In fact, measures found to be reliable in cross-sectional research may not be reliable when used across time within an individual. Prior research has suggested that some constructs, such as *Energy* and *Well-being*, have been shown to have greater consistency in factor solutions across individuals and simpler structures (Lebo & Nesselroade, 1978). More work is needed to arrive at reliable measurement for use when examining individuals across time.

References

Anderson, T.W. (1963). The use of factor analysis in the statistical analysis of multiple time series. *Psychometrika*, *28*(1), 1–25.

Bentler, P.M. (1990). Comparative fit indexes in structural models. *Psychological Bulletin*, *107*(2), 238–246.

Bentler, P.M. & Bonett, D.G. (1980). Significance tests and goodness of fit in the analysis of covariance structures. *Psychological Bulletin*, 88(3), 588–606.

Bollen, K.A. (1989). Structural equation models with observed variables. *Structural Equations with Latent Variables*, 80–150.

Borkenau, P. & Ostendorf, F. (1998). The Big Five as states: How useful is the five-factor model to describe intraindividual variations over time? *Journal of Research in Personality*, *32*(2), 202–221.

Borsboom, D., Mellenbergh, G.J., & Van Heerden, J. (2003). The theoretical status of latent variables. *Psychological Review*, *110*(2), 203.

Brown, T.A. (2006). *Confirmatory factor analysis for applied research*. New York, NY: Guilford Press.

Cattell, R.B. (1943). The description of personality: I. Foundations of trait measurement. *Psychological Review, 50*(6), 559–594.

Cattell, R.B. (1951). P-technique, a new method for analyzing the structure of personal motivation. *Transactions of the New York Academy of Sciences, 14*, 29–34.

Cattell, R.B. & Luborsky, L.B. (1950). P-technique demonstrated as a new clinical method for determining personality and symptom structure. *The Journal of General Psychology, 42*(1), 3–24.

Garfein, A.J. & Smyer, M.A. (1991). P-technique factor analyses of the Multiple Affect Adjective Check List (MAACL). *Journal of Psychopathology and Behavioral Assessment, 13*(2), 155–171.

Hu, L.T. & Bentler, P.M. (1998). Fit indices in covariance structure modeling: Sensitivity to underparameterized model misspecification. *Psychological Methods, 3*(4), 424.

Hu, L.T. & Bentler, P.M. (1999). Cutoff criteria for fit indexes in covariance structure analysis: Conventional criteria versus new alternatives. *Structural Equation Modeling: A Multidisciplinary Journal, 6*(1), 1–55.

Jones, J.J. & Nesselroade, J.R. (1990). Multivariate, replicated, single-subject, repeated measures designs, and P-technique factor analysis: A review of the intraindividual change studies. *Experimental Aging Research, 16*(4), 171–183.

Jones, C.J., Nesselroade, J.R., & Birkel, R.C. (1991). Examination of staffing level effects in the family household: An application of P-technique factor analysis. *Journal of Environmental Psychology, 11*(1), 59–73.

Goodman, S. (1999). Toward evidence-based medical statistics. 1: The p value fallacy. *Annals of Internal Medicine, 130*(12), 995–1004. doi:10.7326/0003-4819-130-12-199906150-00008. PMID 10383371.

Lawley, D.N. & Maxwell, A.E. (1963). *Factor analysis as a statistical method*. London: Butterworths.

Lebo, M.A. & Nesselroade, J.R. (1978). Intraindividual differences dimensions of mood change during pregnancy identified in five P-technique factor analyses. *Journal of Research in Personality, 12*(2), 205–224.

Lo, L.L., Molenaar, P.C., & Rovine, M. (2017). Determining the number of factors in P-technique factor analysis. *Applied Developmental Science, 21*(2), 94–105.

Lord, F.M. & Novick, M.R. (1968). *Statistical theories of mental test scores*. Reading, MA: Addison-Wesley.

Lu, Z.H., Chow, S.M., & Loken, E. (2017). A comparison of Bayesian and frequentist model selection methods for factor analysis models. *Psychological Methods, 22*(2), 361.

Mitteness, L.S. & Nesselroade, J.R. (1987). Attachment in adulthood: Longitudinal investigation of mother-daughter affective interdependencies by P-technique factor analysis. *The Southern Psychologist, 3*(2), 37–44.

Molenaar, P.C.M. & Nesselroade, J.R. (1998). A comparison of pseudo-maximum likelihood and asymptotically distribution-free dynamic factor analysis parameter estimation in fitting covariance-structure models to block-Toeplitz matrices representing single-subject multivariate time series. *Multivariate Behavioral Research, 33*(3), 313–342.

Molenaar, P.C.M., & Nesselroade, J.R. (2009). The recoverability of P-technique factor analysis. *Multivariate Behavioral Research, 44*(1), 130–141.

Raîche, G., Walls, T.A., Magis, D., Riopel, M., & Blais, J.G. (2013). Non-graphical solutions for Cattell's scree test. *Methodology: European Journal of Research Methods for the Behavioral and Social Sciences, 9*(1), 23.

Roberts, M.L. & Nesselroade, J.R. (1986). Intraindividual variability in perceived locus of control in adults: P-technique factor analyses of short-term change. *Journal of Research in Personality, 20*(4), 529–545.

Russell, R.L., Bryant, F.B., & Estrada, A.U. (1996). Confirmatory P-technique analyses of therapist discourse: High- versus low-quality child therapy sessions. *Journal of Consulting and Clinical Psychology, 64*(6), 1366–1376.

Russell, R.L., Jones, M.E., & Miller, S.A. (2007). Core process components in psychotherapy: A synthetic review of P-technique studies. *Psychotherapy Research, 17*(3), 273–291.

Schulenberg, J.E., Vondracek, F.W., & Nesselroade, J.R. (1988). Patterns of short-term changes in individuals' work values: P-technique factor analyses of intraindividual variability. *Multivariate Behavioral Research, 23*(3), 377–395.

Schwarz, G.E. (1978), Estimating the dimension of a model. *Annals of Statistics, 6*(2), 461–464. doi:10.1214/aos/1176344136, MR 468014.

Steiger, J.H. & Lind, J. (1980) Statistically-based tests for the number of common factors. Paper presented at the Annual Spring Meeting of the Psychometric Society, Iowa City.

Tanaka, J.S. (1993). Multifaceted conceptions of fit in structural equation models. In K.A. Bollen & J.S. Long (Eds.), *Testing structural equation models.* Newbury Park, CA: SAGE.

Tucker, L.R. & Lewis, C. (1973). A reliability coefficient for maximum likelihood factor analysis. *Psychometrika, 38*(1), 1–10.

Wooldridge, J.M. (1994). Estimation and inference for dependent processes. In R.F. Engle & D.L. Mcfadden (Eds.), *Handbook of econometrics, Volume IV* (pp. 2639–2738). Amsterdam: Elsevier.

4

Vector Autoregression (VAR)

This chapter covers foundational concepts and draws largely from Lütkepohl (2005), unless otherwise noted. Herein we cover some fundamental topics for time series analysis, beginning with the univariate level before moving to multivariate models. At the univariate level, we discuss concepts such as stationarity, stability, trends, and order selection (i.e., number of lags). We then discuss the full (multivariate) vector autoregression (VAR) model and more advanced concepts such as Granger causality and structural VAR. The topics are presented in a modular fashion so that the reader can return to specific areas as needed. Throughout this chapter, simulated and empirical data examples are provided with references to the corresponding code and data available in the online supplement. As with P-technique, VAR can be conducted in an exploratory framework. We end with a description of current approaches for selecting models from within a VAR perspective, in particular how to select the model order (i.e., number of lags to include in the model). In Chapter 6, we revisit model selection and discuss data-driven approaches for arriving at the pattern of relations among variables in VAR models (for a known lag order).

4.1 Brief Introduction to the Use of AR and VAR Analysis in the Study of Human Dynamics

The term *autoregressive* (AR) describes analysis regarding how a given variable relates to itself at later (or prior) time points. VAR, the multivariate extension, models both AR relations as well as cross-regressive relations, or how a given variable relates to other variables at later and prior time points. Before describing the details of these models, it is important to emphasize that we focus on AR and VAR conducted on time series data. This differs greatly from approaches that examine changes with longitudinal panel data. With time series data, AR estimates can be obtained for each individual. These estimates indicate how the variables are related across time for a given individual. A high positive AR coefficient in the time series context, for instance, indicates that an individual's value at a lag predicts (to some extent) the value at the next time point. When panel data are used in longitudinal studies, the comparisons are made against other individuals. Here, a high positive AR value indicates that an individual's rank ordering (as compared to other individuals) is similar

DOI: 10.1201/9780429172649-4

from assessment on time 1 to time 2. The online supplement for Chapter 1 provides an example of the different inferences made in longitudinal versus intensive longitudinal analyses. To be clear, we only consider intensive longitudinal (time series) analysis of time series data and not longitudinal panel data.

Both AR and VAR models under this definition have been used to investigate various aspects of human dynamics. Conceptually, the AR weight has been described as a measure of *inertia* (Kuppens, Allen, & Sheeber, 2010), or stability in the variable of interest. Low absolute values indicate lower inertia in the corresponding variable across time since randomness or other variables contribute to the changes across time more so than prior values of the same variable. A high positive estimate, by contrast, would suggest that prior values of the variable are highly predictive of later values, evidencing a state of inertia.

When considering relations across time it is important to note that the temporal resolution, or distance between measurements, greatly influences the results. For instance, Kuppens et al. (2010) collected data 10 times a day for two weeks. They found evidence that it might be normative for emotional states to have inertia throughout the day or remain rather stable. Other studies that captured measurements at a lower temporal resolution of only once a day do not find such consistent evidence for inertia when examining moods and symptoms (e.g., Castro-Schilo and Ferrer, 2013; Lane et al., 2019; Wright et al., 2017). Hence, it is important to consider the distance between observations when making inferences regarding the inertia of a given system or construct in individuals.

4.2 Elementary Linear Models for Univariate Stationary Time

Understanding univariate time series models provides a critical foundation for the methods described in this chapter and the remainder of the book. The autocovariance function introduced in Chapter 2 is the feature of primary interest in many research endeavors as it quantifies the sequential dependency of the time series. Since it is so integral to the definitions to follow, we repeat the equation for stationary (i.e., not time-changing) sample autocovariances for a mean-centered time series:

$$c(u) = T^{-1} \sum_{t=u}^{T} \left[y(t) y(t-u) \right], \quad u = 0, 1, \ldots, m \tag{4.1}$$

where $c(u)$ is constant across time for the series and depends only on the lag u and T is the length of observations in the time series. If the autocovariance for a given lag u equals zero, that indicates a lack of dependence at this lag. The correlations are provided by $r_u = c(u)/c(0)$ where $c(0)$ provides the variance, or estimate of $c(u)$ when the lag $u = 0$.

Perhaps the simplest linear time series model is the *first-order AR* model (denoted as AR(1), equivalently "lag one") for a mean-centered, stationary series:

$$y(t) = \phi y(t-1) + \zeta(t), \qquad \forall t. \tag{4.2}$$

The model expressed in (4.2) is called AR because at each time point t the regression is on previous values of the same process. It is called *first order* because the regression of $y(t)$ is only on the previous value $y(t-1)$, not on greater lags (e.g., $y(t-2)$, $y(t-3)$, etc.). The residual time series ζ, a $1 \times T$ vector, is assumed to be a *white noise innovation* series that has zero covariance with all $y(\tau)$, $\tau <= t-1$: $\text{cov}[y(\tau), \zeta(t)] = 0$. In addition, it is assumed that $\zeta(t)$ lacks any sequential dependencies, that is, its autocovariance function is given by

$$\text{cov}\left[\zeta(t), \zeta(t-u)\right] = \delta(u)\psi, \forall u, \tag{4.3}$$

where $\delta(u)$ denotes the Kronecker delta: $\delta(u) = 1$ if $u = 0$ and $\delta(u) = 0$ otherwise. Consequently, the autocovariance function of $\zeta(t)$ only is nonzero at lag $u = 0$, which provides the variance of the residuals.

The AR(1) model given by (4.2) has a simple interpretation: at each time point t, $y(t)$ depends linearly upon $y(t-1)$, where the coefficient ϕ is called the AR coefficient and provides an estimate of the strength of the relation (or extent of continuity) from time $t-1$ to time t. In general $y(t)$ is not completely explained by $\phi y(t-1)$; the unpredictable part is expressed by $\zeta(t)$. Another way to interpret (4.2) is to conceive of $y(t)$ at each time t as consisting of two parts: a part that is transmitted from $t-1$ to t, given by $\phi y(t-1)$, and a part that is new and random at time t, given by the innovation $\zeta(t)$.

The autocovariance function (4.1) associated with the AR(1) is derived as follows. We first determine $\sigma(0)$, the population-level variance at lag $u = 0$:

$$\sigma(0) = \text{cov}\left[y(t), y(t)\right] = \text{cov}\left[\phi y(t-1) + \zeta(t), \phi y(\text{t}-1) + \zeta(t)\right] \tag{4.4a}$$

Expanding (4.4a), making use of $\text{cov}[\zeta(t), \zeta(t)] = \delta(u)\psi = \psi$ (from Equation 4.3) and $\text{cov}[y(\tau), \zeta(t)] = 0$ for $\tau <= \text{t}-1$, yields:

$$\text{cov}\left[\phi y(t-1) + \zeta(t), \phi y(t-1) + \zeta(t)\right] = \phi^2 \text{cov}\left[y(t-1), y(t-1)\right] + \psi \tag{4.4b}$$

Because of the stationarity assumption of the autocovariance function of $y(t)$, it holds that $\text{cov}[y(t-1), y(t-1)] = \text{cov}[y(t), y(t)] = \sigma(0)$. Combining (4.4a) and (4.4b) then yields $\sigma(0) = \phi^2\sigma(0) + \psi$, or

$$\sigma(0) = \psi / \left(1 - \phi^2\right) \tag{4.4c}$$

An important constraint can be deduced from (4.4c). To begin, per the definition of $\sigma(0)$, we know that the variance has to be nonnegative and finite. The variance ψ of the innovation process $\zeta(t)$ also has to be nonnegative and finite. One can see that the variance $\sigma(0)$ will be nonnegative and finite when $(1 - \phi^2)$ is positive and not equal to zero. This implies that ϕ^2 has to be less than 1, or in other words, $\phi^2 < 1$. Hence, the coefficient ϕ in a stationary AR(1) has to obey the following range restriction: $-1 < \phi < 1$.

For lags $u > 0$, we can derive the theoretical expression for the stationary ACF (4.1) of an AR(1) as follows:

$$\begin{aligned}\sigma(u) &= \text{cov}\left[y(t), y(t-u)\right] = \text{cov}\left[\phi y(t-1) + \zeta(t), y(t-u)\right] \\ &= \phi \text{cov}\left[y(t-1), y(t-u)\right] = \phi\sigma(u-1)\end{aligned} \tag{4.5a}$$

In the second part of (4.5a), we have substituted $\phi y(t-1) + \zeta(t)$ for $y(t)$, while in the fourth part use is made of the assumption that $\text{cov}[\zeta(t), y(t-u)] = 0$.

From (4.5a) it follows that $\sigma(1) = \phi\sigma(0)$, where $\sigma(0)$ is given by (4.4c). Also, it straightforwardly follows from (4.5a) that autocovariance at a given lag is

$$\sigma(\text{u}) = \phi^{\text{u}}\sigma(0), u \geq 0. \tag{4.5b}$$

Having the variance of a time series allows for standardization. Often, this is preferred in order to arrive at scale-free estimates of lagged effects. From the autocovariance function, one can obtain an *autocorrelation function* (ACF), which is defined for a stationary process as

$$\rho(\text{u}) = \sigma(u)/\sigma(0) = \phi^{u}, u \geq 0. \tag{4.5c}$$

Because the absolute value of ϕ has to be less than 1 for it to be stationary, it follows that the stationary ACF is a geometrically decreasing function of lag $u > 1$ for an AR(1) process. Given that the ACF also is symmetrical about lag $u = 0$, the ACF of an AR(1) for all lags $u = 0, \pm1, \pm2, \ldots$ is given by

$$\rho(\text{u}) = \phi^{|\text{u}|}, \quad \forall u. \tag{4.5d}$$

The AR(1) model (4.2) can be straightforwardly generalized to an AR(m) model, where m is an integer, $m > 1$, as follows:

$$y(t) = \phi_1 y(t-1) + \phi_2 y(t-2) + \ldots + \phi_{\text{m}} y(t-\text{m}) + \zeta(t), \forall t. \tag{4.6}$$

The derivation of the ACF associated an AR(m), $m > 1$, is similar to, but more complex than the derivation for the AR(1). Also, the conditions to guarantee that (4.6) represents a stationary series are more complex (described below), but similar to the analogous condition ($-1 < \phi < 1$) for an AR(1).

Given the sequential dependencies of many time series data, it is often helpful to assess the unique contributions that a given lag has on subsequent time points. For this, researchers estimate the *partial autocorrelation function* (PACF). Conceptually, the PACF estimates the excess information explained in the time series at no lag, $y(t)$, and itself at a lag, $y(u)$ after accounting for smaller lags.

The PACF, which for a Gaussian stationary series can be written as $\pi(u) = \text{corr}[y(t), y(t-u) \mid y(t-1), y(t-2), \ldots, y(t-u-1)]$.

It can be seen that for an AR(1) process, the estimate for $\pi(1)$ is the autocorrelation at a lag of 1. For lags greater than 1, the partial correlation will be estimated near zero for an AR(1) process. One can check this by using the function:

$$\pi(2) = \frac{\rho(2) - \rho^2(1)}{1 - \rho^2(1)}. \tag{4.7}$$

For a stationary first-order function, $\rho(2) = \rho^2(1)$, $\rho(3) = \rho^3(1)$, and so forth. When $\pi(u)$ is greater than zero that means it explains variance in the variable above and beyond what is explained by prior orders. As an example, for an AR(2) process the autocorrelation at a lag of two is $\rho(2) = \rho^2(1) + \pi(2) - \pi(2)\,\rho^2(1)$. The ACF and PACF for an AR(3) simulated time series with the data-generating model $y(t) = .30y(t-1) - .30y(t-2) - .40y(t-3) + (t)$ is presented in Figure 4.1.

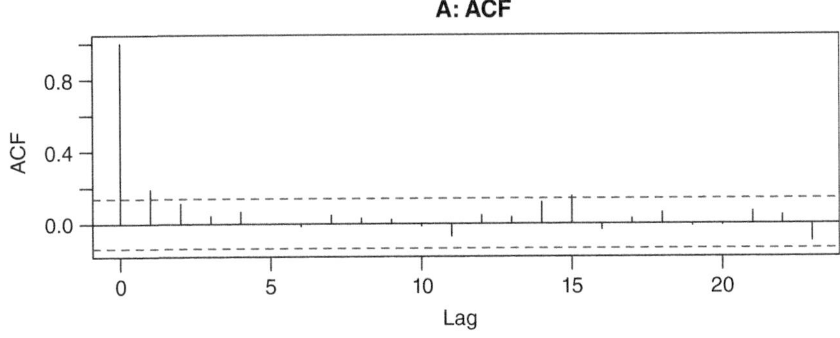

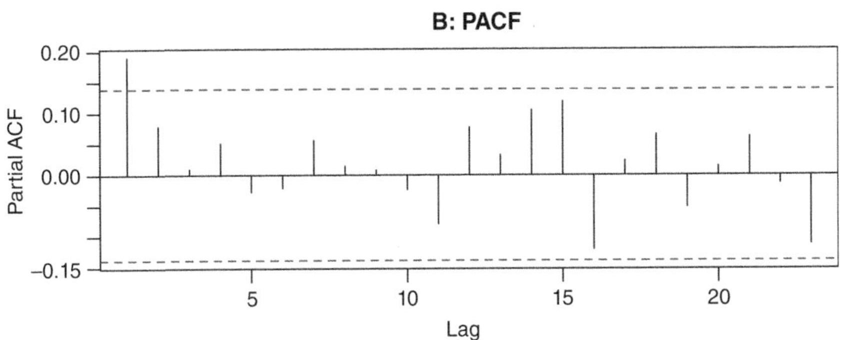

FIGURE 4.1
ACF and PACF for a series generated as an AR(3). $\phi_1 = 0.30$, $\phi_2 = -0.30$, $\phi_3 = -0.40$.

Here, it can be seen that the effect for a lag of 4 is below significance after taking into account the predictive value of lags 1 through 3. Since lags 1–3 are significant, we can ascertain that this is an AR(3) process.

4.3 Stability and Stationarity

A fundamental concept in time series analysis is that of stationarity. Most analytic techniques—including AR and VAR—assume that a univariate time series is stationary. Recall from Chapter 2 that stationary Gaussian processes have constant mean, variance, and covariance estimates across all time points. We begin here with a demonstration of how values can become inflated over time due to the additive nature of time series processes when the AR coefficient is greater than 1. Considering for the moment only a univariate process y, from Equation (4.1) we can estimate the first time point after initialization as

$$y(1) = v + \phi_1 y(0) + \zeta(1) \tag{4.8a}$$

and the second time point as

$$y(2) = v + \phi_1 y(1) + \zeta(2) \tag{4.8b}$$

inserting the right-hand side of Equation (4.7) for y_1 in 4.8, we obtain

$$\begin{aligned} \mathrm{y}(2) &= v + \phi_1 \left[v + \phi_1 y(0) + \zeta(1) \right] + \zeta(2) \\ \Rightarrow y(2) &= (1 + \phi_1) v + \phi_1^2 y(0) + \phi_1 \zeta(1) + \zeta(2) \end{aligned} \tag{4.8c}$$

and iteratively repeating yields

$$y(3) = \left(1 + \phi_1 + \phi_1^2\right) v + \phi_1^3 y_0 + \phi_1 \zeta(1) + \phi_1^2 \zeta(2) + \zeta(3)$$

$$\vdots$$

$$y(t) = \left[\left(1 + \phi_1 + \cdots + \phi_1^{t-1}\right) v + \phi_1^t y(0)\right] + \sum_{u=1}^{t-1} \phi_1^u \zeta(t-u) + \zeta(t) \tag{4.8d}$$

for a given t greater than 3. It is clear to see that a portion of the intercept continues to be added to the right-hand side across time. In the case of non-stationary data, where the coefficient is 1 or larger, this drift (i.e., the intercept) will contribute greatly to each subsequent point due to the continual addition of it to the expected values. Additionally, the multiplication of subsequent errors with the ϕ_1 parameter taken to the power of one less than the current time point indicates that when the absolute value of this parameter is greater than or equal to 1, the error variance will increase across time. Hence, the mean and variance will increase across time, indicating non-stationarity. Also expressed here is the *moving average* (MA) representation of VAR models, meaning that the series can be recast in terms of the innovations and a constant vector. The constant vector here is contained in brackets in Equation (4.8d), with the errors at lags, $\zeta(t-u)$, explaining variability in future observed values of $y(t)$.

In addition to requiring constant mean and covariance structures across time, another desirable feature is that of a *stable* process. Conceptually, a stable process is one that will not grow to infinity (i.e., not "blow up"). Stable processes are always stationary when taken to infinity. For this reason, it is common for researchers to test for stability but actually call it a test for stationarity. Nonetheless, it is important to know that an unstable process can still be stationary (Lütkepohl, 2005, p. 25). An example would be an AR(1) process whereby the AR coefficient is –1 and there is no drift. Here, the values of the observed time series oscillate around a constant mean and variance does not increase. However, when taken to infinity this process would be unstable.

A popular example of a series that violates stationarity both in terms of its mean and autocovariance function and is not stable is the random walk model. At the univariate level, random walks represent a specific type of model that analyzes trends as

$$y(t) = v + y(t-1) + \zeta(t) \tag{4.9}$$

which is initialized with $y_0 = 0$, $\zeta(t)$ is white noise, and v the drift parameter (i.e., intercept), which we saw above can contribute to a change in mean level across time particularly if the AR coefficient is equal to or greater than 1. It is considered an *integrated* series since the value at t is the addition of all previous values plus some error. More specifically, the current values of y are equal to the values of itself at the previous time point plus this drift parameter and the sum of randomness (or shock) introduced by $\zeta(t)$ across time. While not typically included in the formula, this implies that the ϕ_1 estimate is 1. We see immediately that with a coefficient of 1, the model violates the rule implied by 4.4c above, making it a borderline case of non-stationarity. In any case, we can automatically tell given the definition of stability above that this process is not stable.

Figure 4.2A depicts data that were generated to be a random walk with a drift of $v = 0.40$ and an error variance of 5.00. Data for generating these data can be found in the online supplement. In Figures 4.2B and C, we can see that the ACF does not quickly decrease to zero as was evident in the AR(3) example. The PACF indicates that a lag of 1 has a large effect and that all other lags are near zero after taking into account this effect.

Another issue is that data can exhibit changes in mean and variance in a finite sample even if they are stable (and thus stationary) when taken to

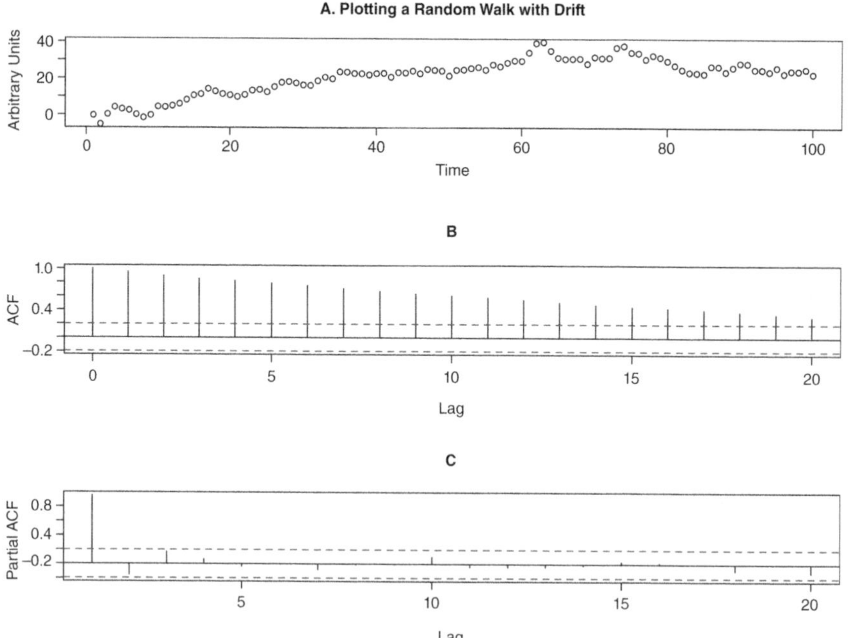

FIGURE 4.2
Depiction of a random walk with drift. The top panel (A) depicts a random walk series generated with a drift of 0.40 and error variance of 5.00. The middle panel (B) presents the ACF for this series, and (C) contains the PACF. Code for generating and plotting the data are available in the online supplement perpetually available through this link: https://www.routledge.com/9781482230598.

infinity. Oftentimes only stability is tested, but changes in variance and mean in finite samples are likely of interest to social, psychological, and neural scientists. As such, tests for both stability and stationarity will be introduced below.

4.3.1 Technical Details Regarding Stability

Mathematically, stability means that the addition of ϕ coefficients times noise taken to infinity ($\Sigma_{u=0}^{\infty}\phi_1^u\zeta_{t-u}$; similar to the finite version seen in the second to last argument of Equation 4.8d) are absolutely summable and a finite solution exists. Absolutely summable simply means the use of absolute values, and a finite solution means the sum of these absolute values does not go to infinity. For an AR(1) process, it is sufficient to identify if the coefficient is greater than 1. For greater lags, we need to use what is called a *characteristic equation* to identify if the system is stable:

$$1-\phi_1 z-\cdots-\phi_m z^m = 0 \quad \text{for } |z|>1. \tag{4.10}$$

Here, m is the max number of lags and z represents solutions for the root. There may be different values of z. If all $|z|$ are greater than 1 when the equation is set to zero, then we know that the roots z are outside the unit circle and the process is stable. If one is less than 1 it is called a *unit root process* and is not stable.

4.3.2 Testing for Stability

Stability tests can easily be implemented using standard software (e.g., the "polyroots" function in R provides this test). Since it is known that a stable process will be stationary, most researchers first test for stability. Most software programs provide easy methods for formally testing if the unit root exists, a criterion for stability being that solutions fall outside the unit root circle. The most common test for stability at the univariate level is the augmented Dickey–Fuller test (ADF). Rejection of the null hypothesis for this test indicates a stable process. Another common test is the Kwiatkowski-Phillips-Schmidt-Shin (KPSS) test (Kwiatkowski et al., 1992). The KPSS test complements the ADF test by identifying if the data that have been identified as unstable via the ADF test will become stable after taking into account a linear trend. Here, the null hypothesis is that the data are stable once the trend is considered, with the alternative hypothesis being that the data are not stable.

4.3.3 Tests for Stationarity

These stability tests have been popularized with economics in mind. In worldwide financial patterns, there can be very large fluctuations or growth in money over time. Psychological, behavioral, social, and neural science researchers likely exhibit very different properties. This is a consequence of the measurements used. For instance, psychophysiological measurements such as heart rate and fMRI data are constrained by what is feasible in a healthy, living person. Similarly, measurements of emotion are bound within a limited range (often a Likert- type scale), and thus will never explode in value even if time is taken to infinity. However, they still may have a trend in the finite sample while being stable. In short time series, heteroscedasticity of variance and trends across time may also appear even though the data still are considered weakly stationary when considered asymptotically (Chatfield, 2003). In fact, it has been argued that solely using unit root tests is a misguided approach for many data formats (Newbold et al., 1993). For these reasons, we now turn to alternative tests of constant mean and variance across time.

Recall that a trend suggests that the value of a given variable varies as a function of time; hence, the expected value or mean changes across time. The most straightforward way to identify if a series contains a trend is to conduct a regression with a trend variable. For a linear trend, this trend variable would simply be a vector, **time** = 1, 2, …*T*. One can simply regress the data on a time vector to arrive at an estimate for the trends:

$$\hat{y}(t) = v + b_{trend} * time(t). \tag{4.11}$$

Should the parameter estimate associated with the variable **time**, b_{trend}, be significantly different from zero, then there is evidence of a trend.

The trend or univariate change across time may either be of interest or simply be a necessary effect to consider when building the model. If the research question pertains to the AR process of a time series, then failing to include the trend might blur the results. Take the data in Figure 4.3 as an example (see online supplement for code). Figure 4.3A ("Data A") contains an AR(1) process: $y(t) = 0.20{*}y(t-1) + \zeta(t)$. The data in Figures 4.3B ("Data B") and 4.3C ("Data C") contain a trend with the same coefficient

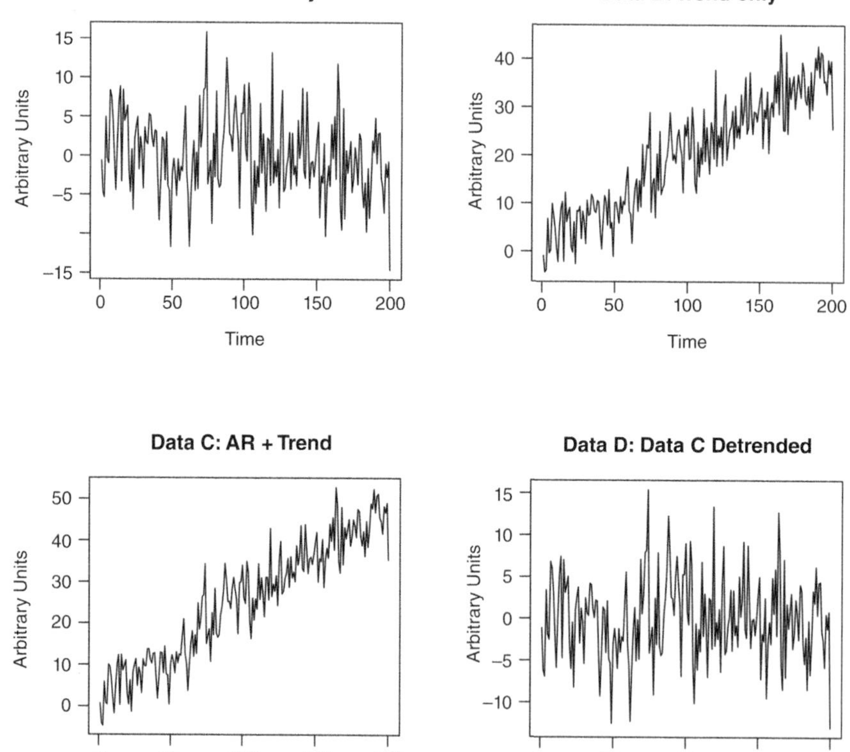

FIGURE 4.3
Depiction of four different simulated time series.

weight (0.20*$time(t)$). However, in Data B there is no AR process. Here, the data are solely explained by the trend and variation: $y(t) = 0.20*time(t) + \zeta(t)$. In Data C, the model contains both the AR(1) process component as well as a trend: $y(t) = 0.20*y(t-1) + 0.20*time(t) + \zeta(t)$. When the parameters in Equation (4.11) are estimated for the data displayed in Figures 4.3A–C, the following trend estimates (i.e., $\hat{b}_{trend}$) and standard errors are, respectively, obtained: 0.00 (se = 0.01), 0.20 (se = 0.01), and 0.25 (se = 0.01). Note that the trend was appropriately identified as present in Data B and C and absent in Data A.

AR(1) estimates are inflated when one does not take into account the trend should one exist. The data-generating parameter for each of the series was $\phi_1 = 0.20$ except for Data B, which had the coefficient $\phi_1 = 0.00$. When AR(1) coefficients are estimated on each of these data examples without accounting for the trend, the following values are obtained: 0.20 (Data A), 0.59 (Data B), and 0.74 (Data C). It is immediately evident that AR(1) coefficients become inflated on account of the linear trend in the data. We discuss in the next section options for modeling or removing trends. For now, having described how to assess if a trend exists, we move onto the detection of changes in variance across time.

Heteroscedasticity, or unequal variance across time, can be assessed with traditional methods used in regression analysis. Specifically, one can assess if the variance of a given y variable varies as a function of the x predictor variable **time**. The Breusch-Pagan test (Breusch & Pagan, 1979) examines the relationship between the squared residuals and the predictor variable. Given our estimated model in Equation (4.11), we can take the squared residuals and regress them on the predictor variable **time** to identify if the variance varies in accordance with the level of the **x** variable (**time**):

$$\zeta(t)^2 = \alpha_0 + \alpha_1 time(t) + \vartheta(t) \tag{4.12}$$

where α_0 and α_1 are intercept and regression coefficients associated with **time**, and $\vartheta(t)$ is the residual. Note that this is a linear function. The coefficient of determination (R^2) from Equation (4.12) times, the sample size N is a test statistic that is chi-square distributed with degrees of freedom equal to the number of predictors. From this, we can statistically test if variance changes as a function of time.

Figure 4.4 depicts a variable that has non-constant variance across time. The data come from a publicly available functional MRI repository collected on control participants and participants diagnosed with Autism Spectrum Disorder (ABIDE; Di Martino et al., 2014).

It can clearly be seen that for the majority of time points in Figure 4.4, the data form a tight cluster around the mean. However, at certain time points, such as from about $t = 350$ to 600, we see large deviations. The large deviations from the mean likely resulted from the individual moving in the scanner.

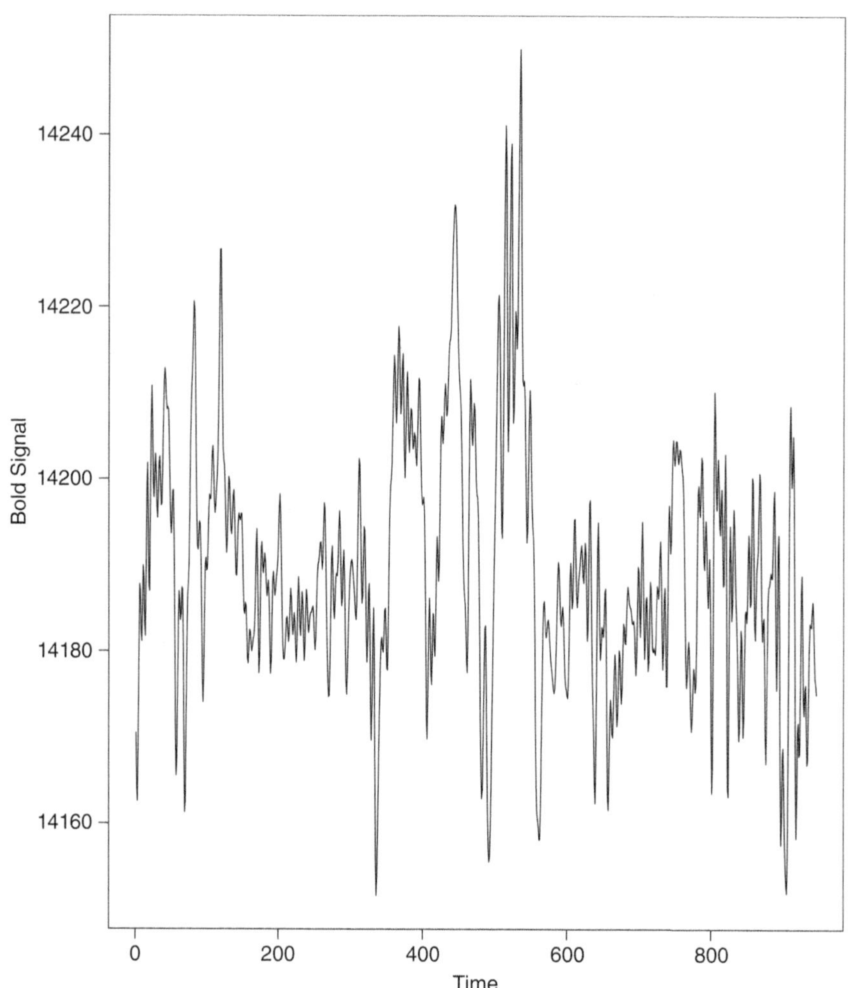

FIGURE 4.4
Example of data with unequal variance across time. Functional MRI blood oxygen level-dependent (BOLD) data were obtained from one subject (see description in Chapter 1) from the ABIDE data (Di Martino et al., 2014).

These data are from a female diagnosed with autism spectrum disorder (ASD), and individuals diagnosed with ASD have an increased propensity for movements as a symptom of this diagnosis. Even slight movement can cause deleterious effects on the data and subsequent inferences drawn from them (Power et al., 2012).[1] In particular, we see great increases in variance

[1] As an aside, in the case of increased variance due to movement in the scanner the deviations do not reflect the true signal at that brain region.

when individuals move during fMRI scanning. A test for heteroscedasticity based on the Breusch-Pagan test, the non-constant variance score test (*ncvTest*) conducted in the R package *car* (Fox, 2022), revealed that these data do in fact differ significantly in the variability as a function of time with a $\chi^2 = 72.78$, df = 1, $p < 0.001$.

Unfortunately, there are no simple transformations that can accommodate this violation of the homoscedasticity assumption. Advanced techniques that are outside the scope of this chapter, such as Autoregressive Conditionally Heteroscedastic (ARCH; Engle, 1982), explicitly model the change in error variance. However, in some cases, the data may violate a test for heteroscedasticity but really it is the mean level that is changing systematically across time. For instance, in the fMRI data in Figure 4.4, the increased variance seen from about $t = 250$ to 600 could be due to a nonlinear effect in the data that changes the level (or mean) via a sharp shift, thereby increasing variance. In fact, nonlinear trends are one common culprit for heteroscedasticity appearing in data. One method for removing nonlinearities is the Box-Cox transformation (Box & Cox, 1964). Here, the optimal power transform is arrived at in a data-driven manner. The difficulty in transformations such as these is that they often render the results difficult to interpret. The optimal approach would be to remove the change in variance in the errors by including a variable that perhaps models this change. In the fMRI example, researchers often include variables that relate to movement to regress out its influence on the data.

4.4 Detrending Data

All of the tests discussed here only assess the univariate level. It is common practice to test for trends and heteroscedasticity at the univariate level prior to conducting multivariate analysis. We now consider how to process data that have been found to have trends, and then move to a discussion of time series models with multiple variables.

There are two general options for handling data with trends: (1) include a trend variable directly in the model or (2) detrend the data prior to analysis. Regarding the first option, many researchers prefer to model the trend effect in the same model as any other effects (such as AR effects, or other variables). A benefit of this approach is that one can immediately see how strong the trend effect is after taking into account the potential influence of other variables and AR effects. For instance, one can see if there is an increase in Positive Affect across time by including the trend, and also identify if other variables (such as Minutes Spent Working Out; prior values of Positive Affect) relate to Positive Affect above and beyond this trend. Additionally, it is possible that the estimate of the trend effect may be inflated due to the absence of other variables that may also relate to increases in the outcome of interest across time. Formally, to include a trend, one would fit the following model for an AR of order m:

$$y(t)=\phi_1 y(t-1)+\phi_2 y(t-2)+\ldots+\phi_m y(t-m)+b_{trend}time(t)+\zeta(t), \forall t. \quad (4.13)$$

Note that this is identical to Equation 4.6 except for the addition of a **time** vector which spans from 1, 2, … T. When an AR(1) version of this model, $y(t) = \phi_1 y(t-1) + b_{trend}time(t) + \zeta(t)$, with **time** being a linear trend is fitted to Data C (which contains both a trend and AR(1) effect) the following estimates are obtained: $\phi_1 = 0.21$, $b_{trend} = 0.20$. These values are close to the 0.20 value used for both coefficients when generating the data.

Note that the vector **time** can take any shape, the researcher hypothesizes might explain variability in the target variable. In fMRI research, for example, the time vector might be the shape of the expected response in brain activity following stimuli presentation. Here, the values of the vector rise and fall throughout the length of $t = 1,2, \ldots T$. It is also possible to look for curvature by including polynomial time variables (e.g., $\mathbf{time}^2$) much like one would in a regular regression framework.

To include a cyclical effect, one could ensure the lag of interest is included. For instance, say we have a process that we believe is an AR(1) process but that it might have weekly effects as well. We could fit the following model:

$$y(t)=\phi_1 y(t-1)+\phi_7 y(t-7)+\zeta(t), \forall t \quad (4.14)$$

to take into account if what occurred, for instance, the last time it was Saturday might influence the present Saturday's value since the behavior or mood can partly be explained by what day of the week it is. Another option would be to include a design vector that matches the expected shape for the weekly effects. This could include a simple indicator variable for the cyclical effect of interest, such as being coded as "1" for days that fall on the weekend.

The second option, detrending the data prior to analysis, is often seen in fMRI data or other situations when the change across time is not of interest. The first of two approaches *linearly detrends* the data and comprises the following steps: (1) estimate the trend effect in the absence of other variables and AR effects (as done above); (2) arrive at the predicted vector ($\hat{y}$) using the estimates obtained in (1); (3) subtract the predicted vector from the observed vector to obtain residuals: $\hat{y}_{detrend} = y - \hat{y}$. These estimated residuals, $\hat{y}_{detrend}$, are the *detrended data*, or data that have had the linear trend removed. Figure 4.3D provides a depiction of the data resulting from linear detrending of Data C in this manner.

When we conduct the AR(1) model $\hat{y}_{detrend}(t) = \phi_1 \hat{y}_{detrend}(t-1) + \zeta(t)$, we obtain the same estimate for the lagged effect (0.21) as when we include the time vector directly in the model (as described in option 1 of this section). For Data C, the AR effect for the previous time point's influence on the next time point is estimated at 0.21. Importantly, regardless of the method used the interpretation of the linear trend is the same and differs from the interpretation of the original data. Specifically, once time is included in the model all inferences must be made in light of the effect of time. In this example, one would say that we found that y has an AR lag 1 effect of 0.21 after considering the change over time in the level of y. Another way to think of the two effects is that the AR effect is modeling the variability in scores that are not predicted by the trend line (and vice versa if trends are included in the same model).

A second way to remove trends from variables is to *difference* the data and then use this differenced data in subsequent analyses. This is commonly done when the data are found to be not stable, such as the case in random walks, and when the trend is not necessarily linear. The technical definition of an integrated series is that there are d unit roots (i.e., roots equal to 1) for the AR coefficients. That is, the coefficients do not fall outside the unit circle, which is a criterion for stability. For a series that is integrated at an order of $d = 1$ (denoted "$I(1)$"), each time point is subtracted from its subsequent time point:

$$\Delta y(t) = y(t) - y(t-1) = w(t). \tag{4.15}$$

Invoking a lagged operator L enables one to write the general form of differencing:

$$\Delta^{\mathrm{d}} y(t) = (1-L)^{\mathrm{d}}\, y(t) \tag{4.16}$$

where greater orders of integration are denoted with a superscript d. One can identify the order of integration by conducing the KPSS test described above, and differencing functions are available in many software packages (such as R, MATLAB, and SPSS). This test will reveal how far back to difference the data in order to obtain a stable data set. These differenced variables can then be used in subsequent analysis. A univariate model called, "Autoregressive Integrated Moving Average (ARIMA)" differences the data during the model estimation. The MA component indicates that the current values are also predicted by the errors in the estimation of the previous time points, as seen in Equation (4.8d).

Interpretation of the differenced data differs from the linearly detrended effects. For ease in the discussion, let's consider the first-differencing case (i.e., integration of order 1, $d = 1$ in 4.15). Each value in the first-differenced variable represents the change in that variable from the previous time point. If it is the same as the prior time point, it would be zero. A larger value indicates an increase, where a smaller value indicates a decrease. The results are thus viewed in terms of change from the previous time point(s). Other predictors that predict the target first-differenced variable would be said to predict the *change* in the target variable. This may be of interest if someone wishes to predict why someone experiences a change in their mood or behavior across days, for instance.

4.5 Univariate Order Selection

Up until now, we have considered the order of the process to be known. Order selection comprises an important step in the analysis. A standard first step is to inspect the ACF and PACF plots that were first introduced in Section 4.1. Figure 4.5

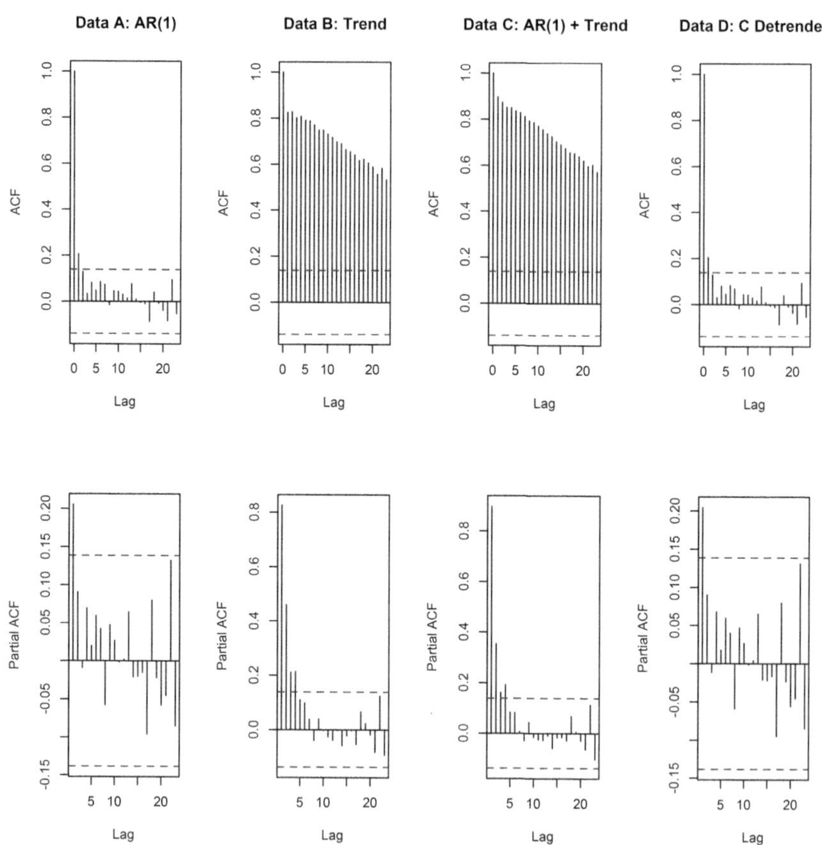

FIGURE 4.5
Depiction of ACF and PACF results for the data displayed in Figure 4.3.

depicts these plots for the data depicted in Figure 4.3 (code is available in the online supplement). The ACF for Data A can be seen to decrease toward zero rapidly after a lag of 1. The PACF also reveals that after taking into account the lag-1 effect, only lag 15 is significant. This often suggests a lag-1 process. Given the multiple tests, it is not uncommon for a spurious effect to emerge, which may explain the coefficient at lag 15 being significant for this AR(1) process.

Inspection of ACF and PACF estimates for time series with trends (Data B and C) quickly reveals that the ACF does not decrease quickly toward zero as would be expected in a stationary process. We also see that the PACF suggests an AR(5) process. Detrending Data C, which contained both an AR(1) effect and trend, returned us to the ACF and PACF plots seen for Data A. Here, the lag is correctly indicated as a lag-1 process. Taken together, this example illustrates two general rules with interpreting these plots with regards to order selection. First, the absence of a quick drop-off to zero for effects in the ACF plot could indicate the presence of some sort of long-term trend. Second, for data that do not contain a trend, the PACF can reveal the optimal lag order for an AR process (Shumway & Stoffer, 2006).

In addition to visual inspection and examination of significance levels of the ACF and PACF estimates, one can evaluate the respective models that would be produced under different lags relative to each other. This is most often done by using a model-selection criterion. Here, we focus on one of the most popular: Aikake information criterion (AIC; Akaike, 1974). The AIC is typically used to examine the relative quality of model fits for different numbers of lags by considering both the average error and the number of coefficients:

$$AIC_{uni}(u) = \ln\left(\frac{RSS_u}{T}\right) + \frac{n+2u}{T}, \tag{4.17}$$

where RSS is the residual sum of squares for the model with a lag order of u. Since the RSS will continue to decrease with more parameters (i.e., lags) added to the model, the AIC penalizes each additional parameter. In this way, the optimal model can be identified that models the data in the most parsimonious way.

Figure 4.6 depicts the difference between the lowest AIC identified (lag = 1) and all other lags for Data A. The models become increasingly lower in quality according to the AIC as more lags are added. An important quality of the AIC is that it does not indicate absolute fit. That is, by not providing a test against a null hypothesis the AIC provides no indication of absolute model quality. All values must be viewed relative to competing models. Here, we see that the best model, or the model with the lowest AIC, is an AR(1), which is the true data-generating lag order.

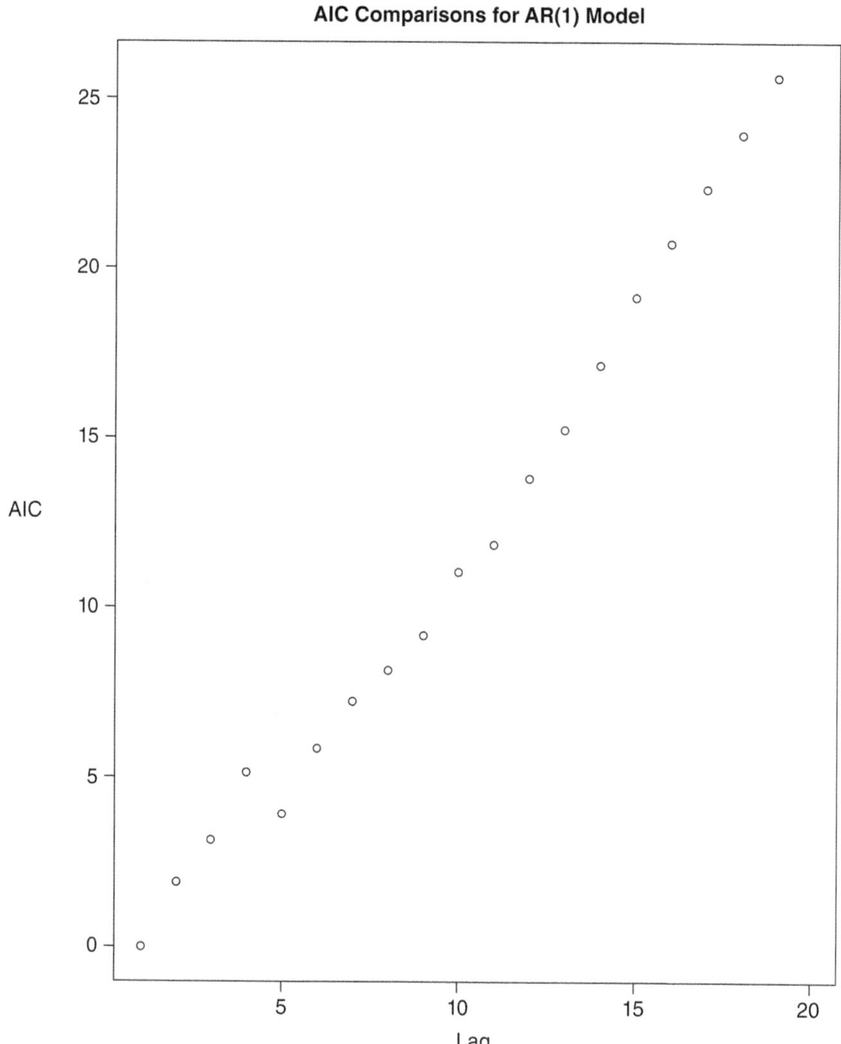

FIGURE 4.6
AIC comparisons for Data A. There is a true lag order of 1 for this data, and the AIC correctly identifies this as the optimal model.

4.6 General VAR Model

Up until now, we have only considered univariate cases. Assessing adherence to assumptions at the univariate level is often conducted prior to multivariate analysis, and thus having a firm understanding of AR processes is a necessary precursor to VAR modeling. We now move to vector (or multivariate) autoregression (VAR). The generalization of the AR(m) for univariate series (4.2) to multivariate series is straightforward. Let $\mathbf{Y}$ be a weakly stationary p-variate observed series. A VAR model with variables lagged up to a finite order m can be represented as

$$\mathbf{y}(t) = \boldsymbol{\upsilon} + \boldsymbol{\Phi}_1\mathbf{y}(t-1) + \boldsymbol{\Phi}_2\mathbf{y}(t-2) + \cdots + \boldsymbol{\Phi}_m\mathbf{y}(t-m) + \boldsymbol{\zeta}(t), \forall t, \tag{4.18}$$

where $\mathbf{y}(t)$ is a p x 1 vector of variables at t, $\boldsymbol{\upsilon}$ is a p x 1 vector of intercept terms, and $\boldsymbol{\zeta}(t)$ is the p x 1 vector of errors in prediction (also referred to as "innovations") at t which are assumed to be noise processes that are identically independently distributed. White noise simply means that the covariance of $\boldsymbol{\zeta}(t)$ with prior or later time points is zero. That is, $\text{cov}[\boldsymbol{\zeta}(t), \boldsymbol{\zeta}(t-u))'] = 0$, with $\text{cov}[\boldsymbol{\zeta}(t), \boldsymbol{\zeta}(t)'] = \boldsymbol{\Sigma}_\zeta$ which is a p x p positive definite symmetric matrix with the error variances on the diagonal. The coefficient matrices $\boldsymbol{\Phi}_u$, with u being equal to 1 through the largest lag of m, are the p x p matrix of coefficients relating the prior time points to the future time point t. The number of lags is indicated in parentheses as VAR(m). For instance, a lag of one would be VAR(1), and one would describe this in words as a VAR model of order 1.

Again the focus will be on the derivation of the mean and covariance function of weakly stationary multivariate time series. Let $\mathbf{Y}$ denote a p-variate time series. That is, $\mathbf{y}(t)' = [y_1(t), y_2(t), \ldots, y_p(t)]$, $t = 1, 2, \ldots, T$, where each component series $\mathbf{y}_k$, $k = 1, 2, \ldots, p$, is a univariate time series. The mean function of the weakly stationary time series $\mathbf{y}_k$ is the constant p-variate vector $\boldsymbol{\mu}_y{}' = [\mu_{y1}, \mu_{y2}, \ldots, \mu_{yp}]$, where μ_{yk} is the constant mean function of $y_k(t)$, $k = 1, 2, \ldots, p$, across all t. The covariance function of $\mathbf{y}(t)$ is defined as

$$\text{cov}\left[\mathbf{y}(t), \mathbf{y}(t-u)'\right] = \text{E}\left[\left\{\mathbf{y}(t) - \boldsymbol{\mu}_y\right\}, \left\{\mathbf{y}(t-u) - \boldsymbol{\mu}_y\right\}'\right] = \Sigma_y(u), u = 0, \pm 1, \ldots. \tag{4.19}$$

Notice that $\text{cov}[\mathbf{y}(t), \mathbf{y}(t-u)']$ is defined as the covariance of a column vector, $\mathbf{y}(t)$, with a row vector, $\mathbf{y}(t-u)'$. This is the so-called outer product of two p-variate vectors, yielding a (p,p)-dimension covariance matrix $\boldsymbol{\Sigma}_y(u)$ at each lag $u = 0, \pm 1, 2, \ldots, m$.

To ease the following presentation, we suppose that $\mathbf{y}(t)$ is a bivariate ($p = 2$) stationary time series. We focus on the autocovariance matrix and take the constant mean function $\boldsymbol{\mu}_y{}' = [\mu_{y1}, \mu_{y2}]$ to be a zero vector. The explicit expression of the covariance function of $\mathbf{y}(t)$ at lag u is

$$\Sigma_y(u) = \begin{bmatrix} \sigma_{11}(u) & \sigma_{12}(u) \\ \sigma_{21}(u) & \sigma_{22}(u) \end{bmatrix}, \quad u = 0, \pm 1, \ldots \tag{4.20}$$

where $\sigma_{kl}(u) = \text{cov}[y_k(t), y_l(t-u)]$, $k, l = 1, 2$. Hence, $\sigma_{11}(u)$ is the autocovariance of variable $\mathbf{y}_1$ at a given lag u and $\sigma_{22}(u)$ is the autocovariance of $\mathbf{y}_2$ at lag u.

This matrix is not symmetric. As such, special care has to be taken in correctly interpreting $\sigma_{12}(u)$ and $\sigma_{21}(u)$.

$\sigma_{12}(u)$ is the covariance between $y_1(t)$ and $y_2(t-u)$. It is commonly referred to as the cross-covariance at lag u between $y_1(t)$ and $y_2(t-u)$. Hence, $\sigma_{12}(u)$ is the cross-covariance between the first series (variable) $\mathbf{y}_1$ and the second series $\mathbf{y}_2$ at lag u. That is, $\sigma_{12}(u)$ pertains to the lagged relationship of the first and second series of **Y** when the *second* series precedes the first component series u time steps. In contrast, $\sigma_{21}(u)$ is the covariance between $\mathbf{y}_2$ at t and $\mathbf{y}_1$ at a lag of u. It commonly is referred to as the cross-covariance at lag u between $y_2(t)$ and $y_1(t-u)$. That is, $\sigma_{21}(u)$ is the cross-covariance between the second component series of **Y** and the first component series of **Y** at a lag of u. Hence, $\sigma_{21}(u)$ pertains to the lagged relationship of the second and first series when the *first* series precedes the second component series u time steps.

In general, $\sigma_{12}(u) \neq \sigma_{21}(u)$ for nonzero lags u. That is, $\sigma_{12}(u)$, the cross-covariance at lag u between $y_1(t)$ and $y_2(t-u)$, differs in general from $\sigma_{21}(u)$, the cross-covariance at lag u between $y_2(t)$ and $y_1(t-u)$. Only at lag $u = 0$ are $\sigma_{12}(0)$ and $\sigma_{21}(0)$ always equal to each other: $\sigma_{12}(0) = \sigma_{21}(0)$. There is, however, a lawful relationship between $\sigma_{12}(u)$ and $\sigma_{21}(u)$ in that (a) $\sigma_{12}(u) = \sigma_{21}(-u)$ and (b) $\sigma_{12}(-u) = \sigma_{21}(u)$.

This can be proven very simply. Remember that $\sigma_{21}(u)$ is the cross-covariance between $y_2(t)$ and $y_1(t-u)$. Hence, $\sigma_{21}(-u)$ is the cross-covariance between $y_2(t)$ and $y_1(t-(-u))$, where of course $y_1(t-(-u)) = y_1(t+u)$. Consequently, $\sigma_{21}(-u)$ pertains to the lagged relationship of the second and first component series of $\mathbf{y}_1$ when the *second* component series $\mathbf{y}_2$ precedes the first component series u time steps. This is exactly the interpretation of $\sigma_{12}(u)$: $\sigma_{12}(u)$ pertains to the lagged relationship of the first and second component series of **Y** when the *second* component series precedes the first component series u time steps (see above). This proves (a): $\sigma_{12}(u) = \sigma_{21}(-u)$. The proof of (b) $\sigma_{12}(-u) = \sigma_{21}(u)$ is similar and therefore will be left to the reader.

The estimate of the covariance function of a p-variate weakly stationary time series $\mathbf{y}(t)$ with zero mean function, observed at $t = 1, 2, \ldots, T$, is

$$\mathrm{C}(u) = \mathrm{T}^{-1} \sum_{t=u,T} \left[\mathbf{y}(t)\mathbf{y}(t-u)'\right], u = 0,1,\ldots,U \tag{4.21}$$

Only $\mathbf{C}(0)$ is a (p,p)-dimensional symmetrical covariance matrix. For lags, $u \neq 0$ $\mathbf{C}(u)$ is, in general, not symmetrical for reasons described in detail above. The estimates for negative lags u are obtained by the relation: $\mathbf{C}(-u) = \mathbf{C}(u)$. Each estimate $c_{kl}(u)$, $k,l = 1, \ldots, p$, is asymptotically unbiased and has asymptotic variance $var[c_{kl}(u)]$ (Hannan, 1970, pp. 209–211).

The correlation function $\Xi_y(u)$ of **Y** is a natural extension of the univariate case and is defined by

$$\Xi_y(u) = \begin{bmatrix} \gamma_{11}(u) & \gamma_{12}(u) \\ \gamma_{21}(u) & \gamma_{22}(u) \end{bmatrix}, u = 0, \pm 1, \ldots \tag{4.22}$$

where $\gamma_{kl}(u) = \mathrm{cor}[y_k(t), y_l(t-u)] = \sigma_{kl}(u)/sqrt[\sigma_{kk}(0)\ \sigma_{ll}(0)]$, $k,l \in \{1,2\}$.

Above we indicated a number of tests and remedial steps for violations of assumptions at the univariate level. It is possible that each variable may have a constant mean and autocovariance function at the univariate level, but differ in its cross-lagged relations with other variables as a function of time. For instance, brain processes might change across the course of a task as one learns the task or becomes habituated. Another example from daily diary studies would be that perhaps the relationship between two symptoms, such as negative affect and risky drinking, may change for an individual as they progress through substance abuse treatment. Hence, identifying consistency across time in the multivariate effects may also be of interest. The majority of methods discussed in this book assume that, in addition to means and autocovariances, the cross-covariances among variables are constant across time. This constituted multivariate stationarity.

There are methods for identifying and accommodating violations of this assumption that will be discussed in Chapter 7. In this case, oftentimes the detection of a violation of this assumption is subsumed in the model-building process. Specifically, one might identify a shift in the multivariate covariance structure using a change point detection technique as a part of the process in regime-shifting modeling (see Baek et al., 2021).

4.7 Multivariate Order Selection

Order selection for a VAR model utilizes the criterion to identify the optimal model. The AIC for the multivariate case is

$$AIC_{multi}(u) = (\ln)\left|\Sigma_{\zeta}(u)\right| + \frac{2up^2}{T}, \tag{4.23}$$

where $\mathbf{\Sigma}_{\zeta}(u)$ is the error covariance matrix for the model estimated with an order u, $|.|$ indicates the determinant, and p is the number of variables. As with the univariate case, one compares the AIC across multiple lags (up to a predefined number) to identify the model with the lowest AIC. Conceptually, this is similar to minimizing the errors while taking into account the number of parameters in the model.

For large values of T, the AIC will at times overestimate the order, although this tends to decrease as the number of variables increases (Paulsen & Tjøstheim, 1985). For this reason, two additional criteria known to be consistent are also used. The Hannan and Quinn (Hannan & Quinn, 1979; Liew, 2004) HQ criterion comprises the following:

$$HQ(u) = \ln \left|\Sigma_{\varsigma}(u)\right| + \frac{\ln\ln T}{T} up^2. \tag{4.24}$$

Schwarz (1978) derived what is commonly referred to as the Bayesian information criterion (BIC) as follows:

$$BIC(u) = \ln \left|\Sigma_{\varsigma}(u)\right| + \frac{\ln T}{T} up^2. \tag{4.25}$$

As with the AIC, the minimum HQ and BIC values indicate the order of the multivariate process. All the criteria discussed here must be considered relative to alternative lags and should not be considered an absolute test of model fit.

To demonstrate these tests, we utilize the data collected by Borkenau and Ostendorf (1989). As described in Chapter 1, these data contain self-report responses to a Big Five personality questionnaire collected across 90 days for 22 individuals. We use here the data for Participant 6 and provide this data along with the code in the online supplement. We selected six variables that are considered to be part of the "conscientiousness" personality trait. These were obstinate, industrious, persistent, responsible, lazy, and reckless. All three criteria suggested a VAR(1) solution. The estimated coefficients for the AR and cross-lagged relations are

$$\boldsymbol{\Phi}_1 = \begin{bmatrix} \mathbf{0.46} & -0.10 & -0.31 & 0.02 & -0.34 & 0.27 \\ -0.21 & \mathbf{0.06} & -0.01 & -0.43 & 0.45 & -0.09 \\ -0.13 & 0.15 & \mathbf{0.17} & 0.09 & 0.07 & -0.28 \\ 0.49 & 0.08 & -0.15 & \mathbf{-0.10} & -0.14 & 0.28 \\ 0.53 & 0.10 & -0.23 & -0.01 & \mathbf{-0.22} & 0.30 \\ -0.50 & -0.06 & 0.18 & 0.08 & 0.24 & \mathbf{-0.16} \end{bmatrix}.$$

The diagonals, bolded for emphasis, represent the AR effects. Following matrix multiplication rules, the matrix can be interpreted on an equation-by-equation level. Row 1, for instance, contains the coefficients for each of the variables' lag-1 relationship with the first variable ("obstinate") after controlling for the other variables. The columns contain the effects estimated for how a given variable predicts the variables (including itself) at a lag-1, or the lag specified for the matrix. For instance, one can see that taking into account the influence of the other variables, previous values of "obstinate" (variable 1) highly predict subsequent values ($\phi_{11} = 0.46$). For "industrious" (variable 2), the partial AR effect is much lower: 0.06. One can also see cross-lagged effects. For instance, we can see that for every unit increase in "lazy" (variable 4) one can expect a decrease in "industrious" (variable 2) the following day after taking into account the other effects.

4.8 Testing of Residuals

Recall that assumptions for both AR and VAR models hold that the errors are white noise. That is, they are not correlated with previous lags. After selecting an order, it is recommended to test the residuals to ensure that these conditions are met. Considering the estimated autocovariances matrices of $\zeta(t)$ as

$$\mathbf{C}_\zeta(u) = \frac{1}{T}\sum_{t=u+1}^{T}\left[\zeta(t)\zeta(t-u)'\right]. \tag{4.26}$$

The autocovariance matrix can be standardized by

$$\Xi_\zeta(u) = \mathbf{F}_\zeta^{-1}\mathbf{C}_\zeta(u)\mathbf{F}_\zeta^{-1} \tag{4.27}$$

where $\mathbf{F}_\zeta$ is a p by p diagonal matrix containing the square roots of the diagonal elements of $\mathbf{C}_\zeta(0)$. Each element $\xi_{\zeta kl}(u)$ of the $\Xi_\zeta(u)$ matrix contains the correlation among errors for two variables k and l at a lag of u. With a large number of observations, $\sqrt{T}\xi_{\zeta kl}(u)$ for $u > 0$ have approximate standard normal distributions (see Lütkepohl, 2005, pp. 159–161, for proof). Hence, the null hypothesis, that the correlation of a variable k with variable l at lag u is zero, can be tested by identifying if the confidence interval arrived at by $\xi_{\zeta kl}(u) \pm 2/\sqrt{T}$ contains zero. This is a popular test, given the ease and intuitive nature. However, one should supplement this test with others since this test might reject the null hypothesis less often than indicated by the significance level chosen for small T. Most commonly, portmanteau tests are conducted.

Portmanteau tests arrive at the overall significance of residual autocorrelations up to lag u. Here, the null hypothesis is

$$H_0 : Q_u = (q_1, \ldots, q_u) = 0 \tag{4.28}$$

which is tested against the alternative hypothesis that Q_u is not equal to zero.

Formally:

$$Q_u = T\sum_{t=1}^{u}\operatorname{tr}\left(\widehat{\mathbf{C}_\zeta}(t)'\,\widehat{\mathbf{C}_\zeta}(0)^{-1}\,\widehat{\mathbf{C}_\zeta}(t)\widehat{\mathbf{C}_\zeta}(0)^{-1}\right) \tag{4.29}$$

Returning to the Borkenau data, we find that a VAR(1) does sufficiently remove sequential dependencies among the data. The asymptotic portmanteau test found a $Q_5 = 157.76$, df = 144, $p = 0.20$.

Since this test has been found to have low power in small samples (Ljung & Box, 1978), the Ljung-Box or Box-Pierce (Box & Pierce, 1970) tests are often used in these cases. The former is

$$Q_{u,k,\,LB} = T(T+2)\sum_{i=1}^{u}\frac{\widehat{\xi_{\zeta k}^{2}}}{T-i} \tag{4.30}$$

with the Box-Pierce test being:

$$Q_{u,k,\,BP} = T\sum_{i=1}^{u}\widehat{\xi_{\zeta k}^{2}}. \tag{4.31}$$

All of these tests are χ^2 distributed with degrees of freedom being $u-m$, where u, in this case, is the lag up to which the autocorrelations of residuals are tested and m is the lag order of the process.

We demonstrate the utility of using the portmanteau tests to ensure the correct order has been selected using the AR(3) data simulation example introduced in the beginning of this chapter. First, we conduct an AR(2) on this data and save the residuals. We find that the Ljung-Box and Box-Pierce tests both reject the null hypothesis that the errors are uncorrelated across time: $Q_{5,\text{LB}} = 46.40$, df = 3, $p < 0.001$; $Q_{5,\text{BP}} = 45.17$, df = 3, $p < 0.001$. When tested on the correct model AR(3), the null hypothesis is retained: $Q_{5,\text{LB}} = 0.63$, df = 3, $p = 0.89$; $Q_{5,\text{BP}} = 0.63$, df = 3, $p = 0.89$. Hence, with an AR(3) model on this data, the errors are indeed uncorrelated across time.

4.9 Structural Vector Autoregression

Equation (4.18) introduced the standard VAR. Recall that an assumption of VAR models is that there may be instantaneous (or contemporaneous) relations among the errors. Another way to model contemporaneous relations is include them as relations among variables in the VAR model. Termed, *structural VAR* (SVAR), the common form is

$$\mathbf{D}\mathbf{y}(t) = \mathbf{\Phi}_1^{*}\mathbf{y}(t-1) + \cdots + \mathbf{\Phi}_{\text{m}}^{*}\mathbf{y}(t-m) + \epsilon(t) \tag{4.32}$$

where $\mathbf{\Phi}_u^{*}$ will have a different estimate than $\mathbf{\Phi}_{\text{u}}$ in a standard VAR if there are correlated (contemporaneous) errors. $\mathbf{D}$ is an asymmetric matrix, with $\mathbf{\Phi}_u^{*} := \mathbf{D}\mathbf{\Phi}_u$ for $u = 1, \ldots, m$ and $\epsilon(t) := \mathbf{D}\boldsymbol{\zeta}(t) \sim (\mathbf{0}, \mathbf{\Sigma}_\epsilon = \mathbf{D}\mathbf{\Sigma}_\zeta\mathbf{D}')$, where $\mathbf{\Sigma}_\zeta$ indicates the covariance matrix of the errors in standard VAR solution. Here, $\boldsymbol{\Sigma}_\epsilon$ will be a diagonal covariance matrix if **D** is selected properly, hence there would be no cross-correlations among the errors at a lag of zero. In order for identification purposes, the diagonal of **D** is often set to 1, and only the lower triangular portion of **D** is estimated to ensure that there are fewer unique estimates than seen in the correlation matrix (i.e., $p(p-1)/2$), but these constraints can be eased in the presence of other constraints in the model.

For estimation purposes, it can be helpful to have the contemporaneous relations on the right-hand side of the equation. This is typical in structural equation modeling and other forms of simultaneous equation estimation. Begin by defining **D** as $(\mathbf{I}_\mathrm{p} - \mathbf{A})$, where $\mathbf{I}_\mathrm{p}$ is a (p,p)-dimensional identify matrix and **A** is a (p,p)-dimensional matrix of contemporaneous coefficients with a zero diagonal, one is able to separate the ones along diagonal in **D** from the coefficients that need to be estimated. This becomes handy since in the case of SEM, the same variables can exist on both the right- and left-hand sides of the equation. Termed *unified SEM* (Kim et al., 2007), with $\mathbf{D} = (\mathbf{I}_\mathrm{p} - \mathbf{A})$ one can transform an SVAR into a model suitable for SEM analyses:

$$\left(\mathbf{I}_\mathrm{p} - \mathbf{A}\right)\mathbf{y}(t) = \boldsymbol{\Phi}_1^* \mathbf{y}(t-1) + \cdots + \boldsymbol{\Phi}_\mathrm{m}^* \mathbf{y}(t-m) + \epsilon(t) \tag{4.33a}$$

$$\Rightarrow \mathbf{y}(t) - \mathbf{A}\mathbf{y}(t) = \boldsymbol{\Phi}_1^* \mathbf{y}(t-1) + \cdots + \boldsymbol{\Phi}_\mathrm{m}^* \mathbf{y}(t-m) + \epsilon(t) \tag{4.33b}$$

$$\Rightarrow \mathbf{y}(t) = \mathbf{A}\mathbf{y}(t) + \boldsymbol{\Phi}_1^* \mathbf{y}(t-1) + \cdots + \boldsymbol{\Phi}_\mathrm{m}^* \mathbf{y}(t-m) + \epsilon(t) \tag{4.33c}$$

That **A** has zeros along the diagonal should now be clear, as a variable cannot be used to predict itself at time. Notably, the errors $\epsilon(t)$ are assumed to be uncorrelated with each other after accounting for the contemporaneous and lagged relations.

One can transform an SVAR into an equivalent standard VAR. The VAR representation will contain correlations in the residuals. Beginning with Equation (4.33a) above, one can return to a standard VAR as follows:

$$\Rightarrow \mathbf{y}(t) = \left(\mathbf{I}_\mathrm{p} - \mathbf{A}\right)^{-1} \boldsymbol{\Phi}_1^* \mathbf{y}(t-1) + \cdots + \left(\mathbf{I}_\mathrm{p} - \mathbf{A}\right)^{-1} \boldsymbol{\Phi}_\mathrm{m}^* \mathbf{y}(t-m) + \left(\mathbf{I}_\mathrm{p} - \mathbf{A}\right)^{-1} \epsilon(t) \tag{4.33d}$$

$$\Rightarrow \mathbf{y}(t) = \mathbf{D}^{-1}\boldsymbol{\Phi}_1^* \mathbf{y}(t-1) + \cdots + \mathbf{D}^{-1}\boldsymbol{\Phi}_\mathrm{m}^* \mathbf{y}(t-m) + \mathbf{D}^{-1}\epsilon(t) \tag{4.33e}$$

$$\Rightarrow \mathbf{y}(t) = \boldsymbol{\Phi}_1 \mathbf{y}(t-1) + \cdots + \boldsymbol{\Phi}_\mathrm{m} \mathbf{y}(t-m) + \zeta(t) \tag{4.33f}$$

From the above-described properties of $\boldsymbol{\phi}_\mathrm{u}^*$ and $\epsilon(t)$, one can return to the matrices obtained from standard VAR: $\mathbf{D}^{-1}\boldsymbol{\Phi}_\mathrm{u}^* = \boldsymbol{\Phi}_\mathrm{u}$ and $\mathbf{D}^{-1}\epsilon(t) = \zeta_\mathrm{t}$. Thus, we can obtain the estimates that would be obtained via standard VAR if one has the matrix estimates obtained in an SVAR. Of course, should contemporaneous relations exist the coefficients and errors of the VAR will be influenced by the values contained in the **D** matrix. If one conducted an analysis on these errors, the contemporaneous relations in **A** can once again be obtained (Gates et al., 2010). Note that should no contemporaneous relations exist, $\mathbf{D}^{-1}$ is simply an identity matrix.

The decision to model the contemporaneous relations via directed relations among observed variables or covariances among residuals depends on the underlying research question and knowledge surrounding the topic being studied. As they are transformations of each other, one cannot say that one approach is "right" while the other is "wrong". That said, often the purpose of time series analysis is to recover the data generating model, or the process that gave rise to fluctuations in the data. Here is where substantive knowledge can help. Modeling relations as covariances among residuals assumes that the contemporaneous relations among the variables are not caused by the variables available in the system of variables available. This assumes a common cause for the two variables that is external to the variables gathered when looking contemporaneously. Modeling contemporaneous relations as covariances is appropriate when the variables are thought to be synonymous, related to a latent construct that is not modeled, or caused by a variable not in the system of variables available. One should model the relations directly among observed variables (as in SVAR) if they believe the variables could causally relate to each other. Please note that while identifying statistical prediction alone does not in itself allow for causal inferences to be made, it can help to understand potential causal influences to be explored. In Chapter 6, we discuss hybrid-SVAR models that allow for both contemporaneous covariances and directed relates, which is useful when the research has no hypothesis regarding the quality of the relations or believes them to be mixed among the variables provided.

4.10 Granger Causality

Granger causality is a statistical test for identifying if a relation should be included in a model. It is often used in tandem with VAR and SVAR to induce sparseness in the coefficient matrices as it helps to ascertain directionality of effects and/or the presence of contemporaneous covariances. Originally introduced within the economics community, Granger causality has also been used on fMRI data and daily self-report data. Tests of Granger causality consider both lagged and instantaneous (i.e., contemporaneous) effects.

Conceptually, Granger causality identifies which cross-lagged or contemporaneous relations exist in a VAR after considering the AR influence for each variable (Granger, 1969). In this way, effects among variables are retained only if a given variable explains variability in subsequent values for a target variable after controlling for the target variable's ability to predict subsequent values of itself. A variable is said to "Granger cause" another variable if these conditions are met.

Granger causality begins by including the AR effects for all variables. As noted above, these weights indicate the degree to which a given variable can be predicted by itself using previous time points. After accounting for these effects, one can identify if a variable Granger causes another variable by looking at the significance levels of the off-diagonal elements of the $\mathbf{\Phi}$ matrices. Formally, one can test the following equations for a set of two variables, y and x:

$$\begin{aligned} y(t) &= v_y + \phi_{y1} y(t-1) + \phi_{yx} x(t-1) + \zeta_{y,t} \\ x(t) &= v_x + \phi_{x1} x(t-1) + \phi_{xy} y(t-1) + \zeta_{x,t} \end{aligned} \tag{4.34}$$

Should the coefficient estimate for ϕ_{yx} be significant, then there is evidence that x Granger causes y. A similar evaluation can be performed on ϕ_{xy}.

The results from the Borkenau and Ostendorf (1989) data presented above can be interpreted from a Granger causality perspective. Focusing on "industrious", we can see that after taking into account the predictive value that industrious has on itself at the next time point, "responsible" and "lazy" have additional influence on the prediction. By contrast, industrious does not offer any additional predictive influence on these variables. Hence, one can conclude that "responsible" and "lazy" Granger cause "industrious". The concept of Granger causality surfaces in many aspects of time series modeling that arise later in the book as it is a fundamental concept for model-building approaches.

4.11 Discussion

This chapter provided insight into fundamental concepts necessary for many types of analysis with multivariate time series data. In particular, the assumptions, diagnostics, and remedies for time series analysis were presented here. Applications to synthetic data demonstrated key concepts, with daily diary and functional MRI data providing motivating examples. This chapter can serve as a reference point for when one is ready for carrying out their own analysis.

The reader should now be primed to continue on to more advanced topics to be covered in this book. A number of topics in this chapter emerge in later chapters. For instance, determining lag order is utilized in Chapter 5, and Granger causality will return in Chapter 6 on model building. Multivariate non-stationarity will again be a focal point when discussing modeling frameworks in Chapter 7 that can capture the changing relations among variables across time. In Chapter 9, we discuss how the output from VAR models can be considered from a network science perspective.

References

Akaike, H. (1974). A new look at the statistical model identification. *IEEE Transactions on Automatic Control, 19*(6), 716–723.

Baek, C., Gates, K.M., Leinwand, B., & Pipiras, V. (2021). Two sample tests for high-dimensional autocovariances. *Computational Statistics & Data Analysis, 153*, 107067.

Borkenau, P. & Ostendorf, F. (1998). The Big Five as states: How useful is the five-factor model to describe intraindividual variations over time? *Journal of Research in Personality, 32*(2), 202–221.

Box, G.E. & Cox, D.R. (1964). An analysis of transformations. *Journal of the Royal Statistical Society: Series B (Methodological), 26*(2), 211–243.

Box, G.E. & Pierce, D.A. (1970). Distribution of residual autocorrelations in autoregressive-integrated moving average time series models. *Journal of the American Statistical Association, 65*(332), 1509–1526.

Breusch, T.S. & Pagan, A.R. (1979). A simple test for heteroscedasticity and random coefficient variation. *Econometrica: Journal of the Econometric Society*, 1287–1294.

Castro-Schilo, L. & Ferrer, E. (2013). Comparison of nomothetic versus idiographic-oriented methods for making predictions about distal outcomes from time series data. *Multivariate Behavioral Research, 48*(2), 175–207.

Chatfield, C. (2003). *The analysis of time series: An introduction*. New York, NY: Chapman and Hall/CRC.

Di Martino, A., Yan, C.-G., Li, Q., Denio, E., Castellanos, F.X., Alaerts, K., & Milham, M.P. (2014). The autism brain imaging data exchange: Towards a large-scale evaluation of the intrinsic brain architecture in autism. *Molecular Psychiatry, 19*(6), 659–557.

Engle, R.F. (1982). Autoregressive conditional heteroscedasticity with estimates of the variance of United Kingdom inflation. *Econometrica, 50*(4), 987–1007. doi: 10.2307/1912773.

Epskamp, S. & Fried, E.I. (2018). A tutorial on regularized partial correlation networks. *Psychological Methods, 23*(4), 617.

Fox, J. (2022). Car: Companion to applied regression. R package version 3.1-0. https://cran.r-project.org/web/packages/car/index.html.

Gates, K.M., Molenaar, P.C., Hillary, F.G., Ram, N., & Rovine, M.J. (2010). Automatic search for fMRI connectivity mapping: An alternative to Granger causality testing using formal equivalences among SEM path modeling, VAR, and unified SEM. *Neuroimage, 50*(3), 1118–1125.

Granger, C.W. (1969). Investigating causal relations by econometric models and cross-spectral methods. *Econometrica: Journal of the Econometric Society, 37*(3), 424–438.

Hannan, E.J. (1970). *Multiple time series*. New York, NY: Wiley.

Hannan, E.J. & Quinn, B.G. (1979). The determination of the order of an autoregression. *Journal of the Royal Statistical Society. Series B (Methodological), 41*(2), 190–195.

Kim, J., Zhu, W., Chang, L., Bentler, P.M., & Ernst, T. (2007). Unified structural equation modeling approach for the analysis of multisubject, multivariate functional MRI data. *Human Brain Mapping, 28*(2), 85–93.

Kuppens, P., Allen, N.B., & Sheeber, L.B. (2010). Emotional inertia and psychological adjustment. *Psychological Science, 21*(7), 984–991.

Kwiatkowski, D., Phillips, P.C.B., Schmidt, P., & Shin, Y. (1992). Testing the null hypothesis of stationarity against the alternative of a unit root. *Journal of Econometrics, 54*(1–3), 159–178. doi: 10.1016/0304-4076(92)90104-Y.

Lane, S.T., Gates, K.M., Pike, H.K., Beltz, A.M., & Wright, A.G. (2019). Uncovering general, shared, and unique temporal patterns in ambulatory assessment data. *Psychological Methods, 24*(1), 54.

Liew, V.K.S. (2004). Which lag length selection criteria should we employ? *Economics Bulletin, 3*(33), 1–9.

Ljung, G.M. & Box, G.E. (1978). On a measure of lack of fit in time series models. *Biometrika, 65*(2), 297–303.

Lütkepohl, H. (2005). *New introduction to multiple time series analysis.* Berlin: Springer Science & Business Media.

Newbold, P., Agiakloglou, C., & Miller, J. (1993). *Long-term inference based on short-term forecasting models* (Vol. 1, pp. 9–25). London: Chapman & Hall.

Paulsen, J. & Tjøstheim, D. (1985). On the estimation of residual variance and order in autoregressive time series. *Journal of the Royal Statistical Society: Series B (Methodological), 47*(2), 216–228.

Power, J.D., Barnes, K.A., Snyder, A.Z., Schlaggar, B.L., & Petersen, S.E. (2012). Spurious but systematic correlations in functional connectivity MRI networks arise from subject motion. *Neuroimage, 59*(3), 2142–2154.

Schwarz, G. (1978). Estimating the dimension of a model. *The Annals of Statistics, 6*(2), 461–464.

Shumway, R.H. & Stoffer, D.S. (2006). *Time series analysis and its applications: With R examples.* Berlin: Springer Science & Business Media.

Wright, A.G., Stepp, S.D., Scott, L.N., Hallquist, M.N., Beeney, J.E., Lazarus, S.A., & Pilkonis, P.A. (2017). The effect of pathological narcissism on interpersonal and affective processes in social interactions. *Journal of Abnormal Psychology, 126*(7), 898.

5

Dynamic Factor Analysis

As described in Chapter 3, P-technique involves the use of traditional factor analysis on multivariate time series obtained from *one* individual. This contrasts the standard use of factor analysis where one conducts analysis using cross-sectional observations obtained for a sample of individuals at one point in time. A drawback of the P-technique approach is that the temporal relations are not directly modeled (Anderson, 1963). This is problematic because lagged relations among variables (either latent or observed) sometimes exist in data collected intensively across time as discussed in Chapter 4 for vector autoregression (VAR). Should the process contain lagged relations, failing to model these may result in the lagged relations surfacing among the errors or unreliable estimates of other parameters. This violates the P-technique assumption that the errors do not contain serial dependencies.

Research questions may also require modeling of temporal dependencies. For instance, one might be interested in investigating if for a given person one latent construct (e.g., depression) increases after an increase of a second latent construct (e.g., anxiety), or if they tend to co-occur (i.e., simultaneously increase or decrease), or if the second construct tends to temporally precede the first. These types of questions necessitate the estimation of latent variables as well as the modeling of temporal dependencies among these latent variables via a VAR perspective. In this chapter, we present Dynamic Factor Analysis, which incorporates VAR modeling into the P-technique framework.

Relations among latent variables across time and their temporal relations with observed variables can be modeled directly in a number of ways. Dynamic factor models (DFMs) represent a class of such methods (Molenaar, 1985). DFMs as we define them in this book explicitly allow for lagged relations both in the *measurement model* in terms of how observed variables relate to latent variables as well as the *structural model* relating the latent variables to each other across time. A number of constraints to this general modeling framework have been introduced, yielding specific model types. First, we present these various DFM approaches, beginning with the general DFM specification before moving to two constrained versions: process factor analysis (PFA) and shock factor analysis (SFA). Next, this chapter presents options for estimation, as well as examples using simulated, daily diary, and fMRI data.

DOI: 10.1201/9780429172649-5

5.1 General Dynamic Factor Models

The DFM is defined by measurement and a structural equation (Molenaar, 1985). The *measurement* component is

$$\mathbf{y}(t) = \sum_{u=0}^{l} \boldsymbol{\Lambda}(u)\boldsymbol{\eta}(t-u) + \boldsymbol{\varepsilon}(t) \tag{5.1}$$

where $\boldsymbol{\Lambda}(u)$, $u = 0, 1, \ldots, l$, is a sequence of (p,q)-dimensional matrices of factor loadings. As in P-technique (Equation 3.1), $\boldsymbol{\eta}(t)$ represents the q-variate latent factor series, and $\boldsymbol{\varepsilon}(t)$ is a p-variate measurement error series with the assumption of no serial dependence or cross-correlation among errors. Note that if $l = 0$ and no other effects are specified beyond those specified in Equation (5.1), one obtains the P-technique model as a special case.

The second equation, referred to in this context as the *structural, state,* or *transition* equation, is

$$\begin{aligned}\boldsymbol{\eta}(t) - \sum_{u=1}^{m} \boldsymbol{\Phi}(u)\boldsymbol{\eta}(t-u) &= \boldsymbol{\zeta}(t) + \sum_{u=1}^{n} \boldsymbol{\theta}(u)\boldsymbol{\zeta}(t-u)\\ \Rightarrow \boldsymbol{\eta}(t) &= \sum_{u=1}^{m} \boldsymbol{\Phi}(u)\boldsymbol{\eta}(t-u) + \boldsymbol{\zeta}(t) + \sum_{u=1}^{n} \boldsymbol{\theta}(u)\boldsymbol{\zeta}(t-u)\end{aligned} \tag{5.2}$$

where, similar to Equation (4.6), $\boldsymbol{\zeta}(t)$ is a q-variate noise process and $\boldsymbol{\Phi}(u)$, $u = 1, 2, \ldots, m$, is a sequence of (q,q)-dimensional auto- and cross-regressive coefficient matrices. The difference here is that we include the moving average component (introduced in Chapter 4), or the sequential relations among equation residuals and latent variables, to represent the full range of structural relations in the form of an autoregressive moving average (ARMA) model. The matrices $\boldsymbol{\theta}(u)$, $u = 1, 2, \ldots, n$, represent the (q,q)-dimensional moving average coefficient matrices reflecting the amount of variability explained in the current latent variable values using residuals from the prediction of previous latent variable values. Given the multivariate nature of DFMs, the structural equation in 5.2 is a vector ARMA (VARMA(m,n)), or a vector autoregressive of order m as well as a moving average model of order n. Taken together with the measurement component, the overall model can be described as DFM(p,q,l,m,n). The length of the lags in the measurement Equation (5.1) is l, and the VARMA components (5.2) indicated by m and n. As before, p and q indicate the number of observed and latent variables, respectively.

It is assumed that $\boldsymbol{\varepsilon}(t)$ and $\boldsymbol{\zeta}(t)$ are Gaussian processes with covariance functions $\text{cov}[\boldsymbol{\varepsilon}(t),\boldsymbol{\varepsilon}(t\text{-}u)] = \text{diag-}\boldsymbol{\Theta}(u)$ for all $u = 0, \pm1, \ldots,$ and $\text{cov}[\boldsymbol{\zeta}(t), \boldsymbol{\zeta}(t)] = \boldsymbol{\Psi}$, and $\text{cov}[\boldsymbol{\zeta}(t), \boldsymbol{\zeta}(t-u)] = 0$ for $u \neq 0$. The assumption regarding the measurement errors $\boldsymbol{\varepsilon}(t)$ indicates that the measurement errors associated with $\mathbf{y}(t)$ can be autocorrelated (i.e., a manifest variable can show sequential dependencies in its own measurement errors over time) but not cross-correlated (i.e., showing sequential dependencies in measurement errors between different manifest variables). Per the second assumption, the process noise $\boldsymbol{\zeta}(t)$ is assumed to lack any sequential dependency. This is identical to the assumption discussed in Chapter 4 for AR and VAR processes.

In the present form, the model is typically not identifiable and requires constraints. Conditions exist for local identifiability with a rule of thumb being that there must be at least q^2 constraints, where q is the dimension of the latent factor series $\boldsymbol{\eta}(t)$. Please note that this rule holds only for factor loadings that are not rotated and when $q > 1$ (Molenaar, 2017). One optional set of constraints would be to require that the residuals from the structural model $\boldsymbol{\zeta}(t)$ are uncorrelated with each other. If scaled, the covariance matrix of $\boldsymbol{\zeta}(t)$ is the identity matrix: $\text{cov}[\boldsymbol{\zeta}(t), \boldsymbol{\zeta}(t-u)] = \boldsymbol{\delta}(u)\boldsymbol{\Psi} = \mathbf{I}$ (implying $q^*(q + 1)/2$ constraints; Bai & Wang, 2012). In this case, Bai and Wang (2012) also constrained the zero-order matrix of factor loadings ($\boldsymbol{\Lambda}(0)$) such that only the lowest triangular entries are freely estimated. Note that these minimal local identifiability constraints are independent of the maximum lag l of the factor loadings and the orders m and n of the VARMA for $\boldsymbol{\eta}(t)$. Interested readers are referred to Bai and Wang (2012, 2014) for further details. We now turn to common special cases of this general model which inherently have constraints allowing for tractable models.

5.1.1 Process Factor Analysis

PFA models contain lagged relations at the latent level but do not include lags within the measurement model (see Figure 5.1A). PFA models are also referred to as *direct autoregressive factor models* (Ferrer & Nesselroade, 2003; Nesselroade et al., 2002). The model also coincides with one possible form of linear *state-space models*, which similarly models observations as depending on some latent state, which is driven by a stochastic process (discussed in detail later in this chapter). Regardless of the nomenclature used, these models contain both the *measurement* component relating the observed variables to latent constructs contemporaneously and a *structural* component that relates the latent constructs to their lagged counterparts. Whereas the general DFM approach has lagged relations within the measurement

A. Process Factor Analysis

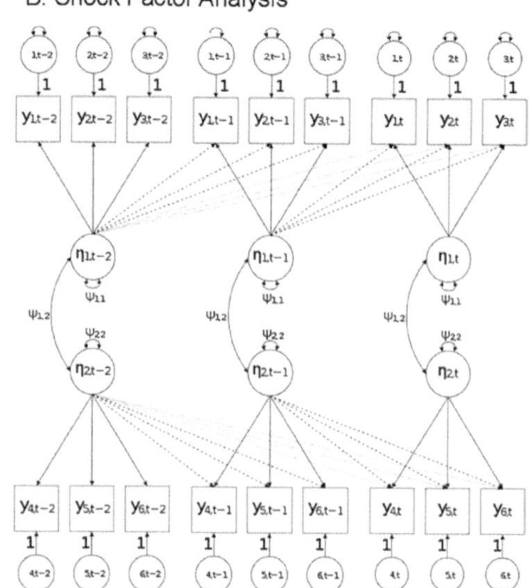

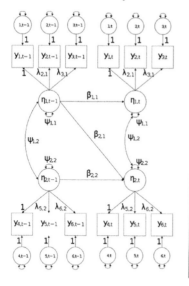

FIGURE 5.1
Depiction of (A) an PFA(1,0) and (B) an SFA(2)model. Factor loadings depicted as dashed arrows in (B) are lag-1 loading parameters that are constrained to be equal for lags with similar dash patterns; in both (A) and (B), factor loadings depicted as solid arrows reflect contemporaneous loadings constrained to be equal for all time points.

model, measurement models are constrained in PFAs as in traditional P-technique:

$$\mathbf{y}(t) = \mathbf{\Lambda}(0)\boldsymbol{\eta}(t) + \boldsymbol{\varepsilon}(t) \tag{5.3}$$

where, as before, $\mathbf{y}(t)$ indicates a p-dimensional observed series at the time, $\mathbf{\Lambda}(0)$ is the (p,q)-dimensional matrix containing the loadings for each of the p observed variables onto the q latent variables contained in $\boldsymbol{\eta}(t)$, and $\boldsymbol{\varepsilon}(t)$ is a vector of measurement errors at time t. This measurement equation reflects the contemporaneous relations between a latent construct and observed variables. $\mathbf{\Lambda}(0)$ is taken as constant such that at any lag, one would expect the same contemporaneous (lag-0) relations among latent and observed variables hold.

For PFA models, the structural equation relating the latent variables across time often does not include the moving average component in 5.2. The model then reduces to a VAR(m) model of the latent variable vector $\boldsymbol{\eta}(t)$:

$$\boldsymbol{\eta}(t)=\sum_{u=1}^{m}\boldsymbol{\Phi}(u)\boldsymbol{\eta}(t-u)+\boldsymbol{\zeta}(t) \tag{5.4}$$

where each (q,q)-dimensional matrix $\boldsymbol{\Phi}(1)$, $\boldsymbol{\Phi}(2)$, …, $\boldsymbol{\Phi}(m)$ contains the regression coefficient estimates for the prediction of a $(q,1)$-dimensional vector of latent variables at time ($\boldsymbol{\eta}(t)$) using previous values of latent variables $\boldsymbol{\eta}(t\text{-}u)$ up to a given lag u = 1, …, m. The diagonals of a given $\boldsymbol{\Phi}(\boldsymbol{u})$ matrix contain the autoregressive (AR) estimates at that given lag, or the estimate for magnitude at which a latent variable predicts itself at later time points. The off-diagonal elements contain estimates of the cross-lagged effects, or the degree to which other latent variables predict a given latent variable at a later time (up to the specified lag m). The measurement errors are typically assumed to have no sequential dependencies: $\text{cov}[\boldsymbol{\varepsilon}(t),\boldsymbol{\varepsilon}(t\text{-}u)] = \delta(u)\boldsymbol{\Theta}$, where $\delta(u)$=0 when $u > 0$ and 1 when there is no lag ($u = 0$), but this condition can be relaxed. As with P-technique, $\boldsymbol{\Theta}$ is assumed to be diagonal.

To describe a PFA model, one often writes PFA(m,n) to indicate the number of lags on the structural model and whether it is a VAR or VARMA. PFA(m,0), or a VAR(m) on the latent variables, is the most commonly used model in the literature. This is equivalent to writing DFM(p,q,0,m,0). Figure 5.1A depicts a PFA(1,0) model with two latent variables and one cross-lagged relationship among these variables.

A number of constraints can be introduced within the PFA framework. For instance, one could arrive at an observed variable VAR model (introduced in Chapter 4) by setting $q = p$ latent variables to be equal to the p observed variables. This can be done by setting the diagonal of the (p,p)-dimensional $\boldsymbol{\Lambda}(0)$ matrix to one, and forcing the measurement errors to be zero. Then, the relations among variables represent the relations among observed variables across time. One can also constrain the model such that no lagged relations exist in the structural model; in this case, one returns to the P-technique factor model. Contemporaneous relations can also be added to this model by allowing for $u = 0$ in Equation (5.4). This would provide a version of PFA similar to a uSEM or SVAR as defined for observed variables in Chapter 4.

The PFA represents the most commonly used DFM in many areas of the social sciences. However, as discussed in Molenaar (2017), this approach may not always reflect the actual data generating processes by neglecting to include lagged factor loadings. An example of a process that may include lagged factor loadings might be a situation where specific symptoms (e.g., anhedonia and fatigue) of the latent construct depression are influenced by the prior day's level of depression, and this may be in addition to the influence of the current day's depression on those symptoms. Importantly, by not investigating the potential for lagged factor loadings the model may miss important relations. We now turn to an approach that allows lagged factor loadings.

5.1.2 Shock Factor Analysis

SFA, which is also referred to as the "white noise factor score model" (Molenaar, 1985), directly captures the lagged relations between observed variables and latent variables. SFA utilizes only Equation (5.1) above, the measurement component of DFM. Conceptually, the factors are considered to be random shocks that are distributed independently across time, but may be contemporaneously related. Browne and Nesselroade (2005) used "daily hassles" as one example of what these latent shocks may represent in empirical scenarios. Here, daily hassles constitute shocks that may affect an individual's responses to the p observed variables at the current as well as some future time points; but daily hassles in and of themselves do not show serial dependencies over time.

The (p,q)-dimensional matrices $\Lambda(0)$, $\Lambda(1)$, …, $\Lambda(l)$ represent the effects of shock (latent) vector variables $\eta(t)$ prediction error decomposition function (l) on the manifest vector variables at the same time point ($\mathbf{y}(t\text{-}l)$), one time point in advance ($\mathbf{y}(t\text{-}l\text{+}1)$), and so on until $\mathbf{y}(t)$. $\Lambda(0)$ is referred to as contemporaneous factor loading matrix, whereas $\Lambda(1)$, …, $\Lambda(l)$, are lagged factor loading matrices. The residual variance-covariance $\Psi(u)$ matrix of the latent shock variables at each lag can be full, with $\text{cov}[\zeta(t), \zeta(t-u)] = 0$ for $u \neq 0$, a set of constraints suggested by Nesselroade and colleagues (e.g., Browne & Nesselroade, 2005; Ferrer & Nesselroade, 2003).

Figure 5.1B depicts this process with contemporaneous covariation among the shock variables. A lag of 2 is depicted here. As compared to PFA, we see it has additional latent (shock) variables and removes the lagged relations among the latent variables. An additional benefit of the SFA model is that relations among observed time series can be modeled (not pictured here). One could further constrain the covariance matrix such that, in the case of $q > 1$, the shock variables do not covary with each other (as in Molenaar, 1985), but this would only be suggested if there is a reason to believe that no relation exists among the q underlying constructs. As can be seen, a given observation at time t is influenced both by the latent construct at time t as well as by earlier values (e.g., $t-1$ and $t-2$) of the latent construct. An SFA(l) indicates an SFA of order l, which is equivalent to writing DFM(p,q,l,0,0).

One way to interpret the relations in SFA is that a $\Lambda(u)$ has been estimated for the observed variables' relationships to latent constructs at previous time points. The idea here is that the lagged construct has delayed influence on observed variables above and beyond contemporaneous influence. As provided as an example above, feelings of depression on a prior day may influence observed indicator values on the next day. Future observed variables' relations to a given latent variable take into account the influence of these earlier latent variables. In this way, it models shocks to the system, or changes in the observed variables that are not accounted for by the static measurement model. The emphasis is on the latent shocks and lasting effects of latent constructs that drive a manifest series.

5.2 Lag Order Selection

The DFM requires that the researcher indicate the following: (1) the number of latent variables q; (2) the number of lags for the measurement model l; and (3) the number of lags for the structural model m and n. Frequently, the researcher only knows for certain p, the number of observed variables. As done in P-technique and VAR, one can select both of these by comparing a set of Candidate Models (CMs) with varying values for q, l, m, and n.

To start, if one does not have any preconceived notions on the preferred form of the DFM(p,q,l,m,n)—in other words, no existing theoretical beliefs to indicate whether an SFA or a PFA model may be preferred—one strategy is to begin with analysis where only q and l vary in a series of P-technique and SFA analyses. Once the number of latent variables q is identified, the researcher can then conduct a PFA of varying lag orders to determine what order of m and n best describes the data or an SFA model at varying values of l to determine the optimal l. In cases where the researcher wants to rely on model fit measures to select between a PFA model and an SFA model, the researcher can compare the fit of the optimal DFM(p,q,0,m,n) and DFM(p,q,l,0,0) models, (or PFA(m,n) and SFA(l), respectively). In other scenarios, the researcher may have reason to use simpler models (e.g., an SFA(0) or P-technique model) as the basis for exploring the optimal q in the first step, and then finding the order of the VARMA in a second step using the factor score estimates. The precise sequence of exploratory steps may vary from application to application, and likely require some adaptations on the part of the researcher to better address the research questions of interest. Below we provide an example of model selection from within an SEM framework. We also provide details on estimation approaches.

5.3 Estimation

Regardless of the model chosen—for example, P-technique, PFA, or SFA—there are three primary estimation techniques within the frequentist approach[1]: the Block–Toeplitz approaches used in the context of SEM often conduct model estimation via quasi-Maximum Likelihood (quasi-ML) and Model-Implied Instrumental Variable-Two Stage Least Squares (MIIV-2SLS),

[1] Bayesian approaches for the estimation of time series models exist but are not covered here as this book aims to provide an overview of techniques that are most common for ecological momentary assessment data.

and raw data likelihood approach based on Kalman filtering (KF). Each of these estimation techniques is demonstrated here using simulated data. The measurement component for this simulated data is similar to the structure and values described in Chapter 3:

$$\boldsymbol{\Lambda}(0)=\begin{bmatrix} 1 & 0 \\ 2 & 0 \\ 1 & 0 \\ 0 & 1 \\ 0 & 2 \\ 0 & 1 \end{bmatrix} \quad \boldsymbol{\Psi}(0)=\begin{bmatrix} 2.77 & \\ 2.47 & 8.40 \end{bmatrix} \tag{5.5}$$

However, for the present data examples, there are AR effects for both latent variables ($\phi(1)(1,1)$ and $\phi(1)(2,2)$) and a cross-lagged effect for $\eta_1(t-1)$ predicting $\eta_2(t)$:

$$\boldsymbol{\Phi}(1)=\begin{matrix} 0.50 & 0 \\ 0.40 & 0.50 \end{matrix} \tag{5.6}$$

The measurement model residuals are uncorrelated across time and with each other. Hence, the data are generated to conform to a PFA(1,0) model. The time series is of length $T = 1000$. Code for simulating the data is available in the online supplement to the book. For ease in discussion, we focus on this model when describing the estimation methods, but do provide an SFA model example. All analyses can be replicated using code in the online supplement.

5.3.1 SEM Estimation with Maximum Likelihood

The data analysis examples previously used in Chapter 3 used traditional maximum likelihood (ML) estimation of SEM. Given that the rows of the data are not independent (explained in detail below) and this violates an assumption of ML, the estimation is best referred to as "quasi" -ML. SEM estimated via quasi-ML is among the most commonly used approaches within social sciences for estimating these time series models, largely because it is a familiar and flexible approximation to DFA that has performed well in simulation studies (Hamaker & Dolan, 2009). Furthermore, the SEM-based approach enables data-driven searches and model building. Utilizing SEM for estimation requires generating a block-Toeplitz matrix. This is done by first time embedding the raw data, which involves copying the data, removing the first row, and pasting it next to the original data, as illustrated in Table 5.1.

One can see from this example that the values of a given variable at $t-1$ are the same as the values of the same variable at t in the prior row. We have highlighted a value for y_1 as an example. This new matrix can be referred to as $\mathbf{Y}_{\text{Toep}}$.

TABLE 5.1

Example of Time Embedded Data

y1(t−1)	*y2(t−1)*	*y3(t−1)*	*y4(t−1)*	*y5(t−1)*	*y6(t−1)*	*y1(t)*	*y2(t)*	*y3(t)*	*y4(t)*	*y5(t)*	*y6(t)*
1.95	2.89	1.64	−2.70	−3.10	−1.23	−1.35	−2.68	−1.60	−6.39	−12.41	−7.98
−1.35	−2.68	−1.60	−6.39	−12.41	−7.98	−1.19	−2.33	−2.20	−7.20	−13.96	−6.85
−1.19	−2.33	−2.20	−7.20	−13.96	−6.85	0.51	−1.92	−1.39	−2.85	−5.10	−3.01
0.51	−1.92	−1.39	−2.85	−5.10	−3.01	−1.89	−4.11	−2.66	−1.77	−5.09	−2.48
−1.89	−4.11	−2.66	−1.77	−5.09	−2.48	−3.15	−4.04	−2.47	−2.20	−5.75	−2.87

In this way, the first set of p variables are the variables at a lag (t – 1), and the next set of p variables are at the next time point t. For a PFA(m,0), the measurement model becomes:

$$\mathbf{y}\left(t-m\|\ldots t-1\|t\right)=\mathbf{\Lambda}_{\mathbf{Toep}}\boldsymbol{\eta}\left(t-m\|\ldots t-1\|t\right)+\boldsymbol{\varepsilon}\left(t-m\|\ldots t-1\|t\right) \tag{5.7}$$

where the symbol || indicates concatenation; hence, the $\mathbf{y}$, $\boldsymbol{\eta}$, and $\boldsymbol{\varepsilon}$ vectors are no longer just the p variables at t but also the p variables at each t-m, …, t – 1 appended before them.

The subscript "Toep" indicates a matrix adjusted to accommodate the lags, with $\mathbf{\Lambda}_{\text{Toep}}$ being a $(m+1)p$ by $(m+1)q$-dimensioned matrix. For a PFA(1,0) with $p = 6$ and $q = 2$ becomes

$$\mathbf{\Lambda}_{\text{Toep}}=\begin{bmatrix} \overbrace{\lambda_{110}}^{\eta_1(t-1)} & \overbrace{0}^{\eta_2(t-1)} & \overbrace{0}^{\eta_1(t)} & \overbrace{0}^{\eta_2(t)} \\ \lambda_{210} & 0 & 0 & 0 \\ \lambda_{310} & 0 & 0 & 0 \\ 0 & \lambda_{420} & 0 & 0 \\ 0 & \lambda_{520} & 0 & 0 \\ 0 & \lambda_{620} & 0 & 0 \\ 0 & 0 & \lambda_{110} & 0 \\ 0 & 0 & \lambda_{210} & 0 \\ 0 & 0 & \lambda_{310} & 0 \\ 0 & 0 & 0 & \lambda_{420} \\ 0 & 0 & 0 & \lambda_{520} \\ 0 & 0 & 0 & \lambda_{620} \end{bmatrix} \tag{5.8}$$

where λ_{210} indicates the parameter relating latent variable 1 to observed variable 2 at no lag, and equivalent subscripts indicate values constrained to be equal. The first $p = 6$ rows correspond to $\mathbf{y}(t$ – 1) with the second 6 rows corresponding to $\mathbf{y}(t)$. Hence, the lagged variables at t – 1 have the same measurement model and estimates as the variables at t.

The latent variable model equation becomes

$$\boldsymbol{\eta}\left(t-m\|\ldots t-1\|t\right)=\boldsymbol{\Phi}_{\text{Toep}}\boldsymbol{\eta}\left(t-m\|\ldots t-1\|t\right)+\boldsymbol{\zeta}\left(t-m\|\ldots t-1\|t\right) \quad (5.9)$$

where the vector definitions are as before. The matrix $\boldsymbol{\Phi}_{\text{Toep}}$ is $(m + 1)q$ by $(m + 1)q$-dimensioned with the columns being independent variables and the rows the dependent variables. The m $\boldsymbol{\Phi}(u)$ matrices occur along the bottom of the $\boldsymbol{\Phi}_{\text{Toep}}$ matrix as the only dependent variables are $\boldsymbol{\eta}(t)$, or the final set of q variables. That is, prior lags are not dependent variables but are included in the model only so they can be predictors. Thus, the variables with the top mq rows are constrained to zero. For a PFA(1,0), the $\boldsymbol{\Phi}_{\text{Toep}}$ matrix would be as follows:

$$\boldsymbol{\Phi}_{\textbf{Toep}}=\underbrace{\begin{bmatrix} 0 & \cdots & \cdots & \cdots & 0 \\ \vdots & \ddots & & & \vdots \\ \vdots & & \ddots & & \vdots \\ \vdots & & & \ddots & \vdots \\ 0 & \cdots & \cdots & \cdots & 0 \\ \phi_{11} & \cdots & \cdots & \cdots & \phi_{1p} \\ \vdots & \ddots & & & \vdots \\ \vdots & & \ddots & & \vdots \\ \vdots & & & \ddots & \vdots \\ \phi_{p1} & \cdots & \cdots & \cdots & \phi_{pp} \end{bmatrix}}_{Lag1}\!\!\underbrace{\begin{bmatrix} 0 & \cdots & \cdots & \cdots & 0 \\ \vdots & \ddots & & & \vdots \\ \vdots & & \ddots & & \vdots \\ \vdots & & & \ddots & \vdots \\ 0 & \cdots & \cdots & \cdots & 0 \\ 0 & \cdots & \cdots & \cdots & 0 \\ \vdots & \ddots & & & \vdots \\ \vdots & & \ddots & & \vdots \\ \vdots & & & \ddots & 0 \\ 0 & \cdots & \cdots & \cdots & 0 \end{bmatrix}}_{Contemporaneous} \quad (5.10)$$

Note the lower left-hand corner contains $\boldsymbol{\Phi}(1)$.[2]

The $(m + 1)p$ by $(m + 1)p$ covariance matrix, $\mathbf{C}_{\textbf{Toep}}$, of the time-embedded $(T\text{-}m)$ x $(m+1)p$ matrix of observed variables provides the data input for quasi-ML estimation from within an SEM framework. The block-Toeplitz approach involves reformulating measurement models from time series data as SEMs. The estimation can be conducted using traditional software that has SEM estimation capabilities. The following model is fit:

$$\mathbf{C}_{\text{Toep}}=\boldsymbol{\Lambda}_{\text{Toep}}\ (\mathbf{I}-\boldsymbol{\Phi}_{\text{Toep}})^{-1}\boldsymbol{\psi}_{\text{Toep}}\ (\mathbf{I}-\boldsymbol{\Phi}_{\text{Toep}}{}')^{-1}\boldsymbol{\Lambda}_{\text{Toep}}+\boldsymbol{\Theta}_{\text{Toep}} \quad (5.11)$$

where and $\boldsymbol{\Phi}_{\text{Toep}}$ the $((m + 1)q$, $(m + 1)q)$-dimensional beta matrix containing estimates for directed (i.e., not correlational) relations among latent variables. $\boldsymbol{\Theta}_{\text{Toep}}$ is the $((m+1)p$, $(m+1)p)$-dimensional covariance matrix of the measurement errors: $\boldsymbol{\Theta}(u) = \text{cov}[\boldsymbol{\varepsilon}, \boldsymbol{\varepsilon}']$ which is diagonal (for reasons explained in Chapter 3) at a given lag and $\boldsymbol{\Lambda}_{\text{Toep}}$ contains the factor loadings. When the latent variable model relations are modeled in $\boldsymbol{\Phi}_{\text{Toep}}$, $\boldsymbol{\Psi}_{\text{Toep}}$ contains the variance/covariance matrix of the *residuals* of the latent variables factor scores for those latent variables that are endogenous $(\text{cov}[\boldsymbol{\zeta}, \boldsymbol{\zeta}'])$ as well as the variance/covariance matrix of the latent variables themselves for those that are not predicted by other latent variables.

[2] Should we opt to conduct a uSEM (see Chapter 4), the lower right-hand corner would also be eligible for estimation. These would be $\boldsymbol{\Phi}(0)$ relations, which is commonly referred to as the **A** matrix.

For a PFA(1,0), the (q,q)-dimensional residual variance/covariance matrix is found in the lower right-hand corner of the full $\boldsymbol{\Psi}_{\text{Toep}}$ matrix, whereas the (q,q)-dimensional latent variable variance/covariance matrix is the upper left corner:

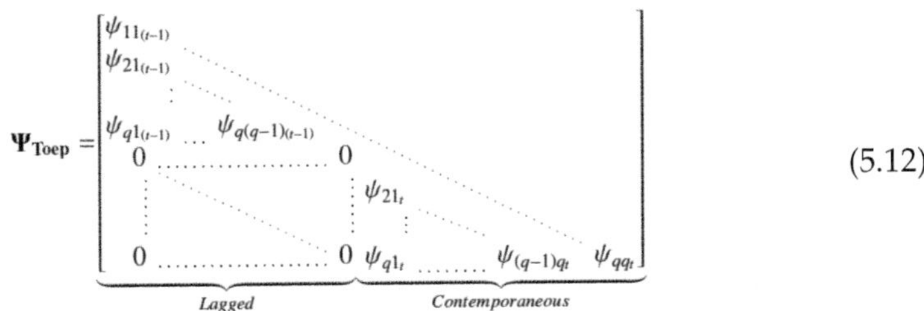

(5.12)

Note that this latter variance/covariance matrix is identical to the interpretation described in Chapter 3. When only the measurement component is modeled (as in the lagged variables), $\boldsymbol{\Psi}$ is the covariance matrix of the factor scores. The lower right-hand matrix represents the variance/covariance matrix among latent variables that could be endogenous (i.e., those at t that are explained by other latent variables at $t-1$). Hence, this is the variance/covariance matrix of residuals $\boldsymbol{\zeta}(t)$. The process noise for the predicted latent variables at t also does not correlate with the latent variables at prior time points—these relations are captured in the directed $\boldsymbol{\Phi}_{\text{Toep}}$ matrix, not in this bidirectional matrix. For this reason, the lower left-hand corner is set to zeros. Concrete examples of elements contained in each of these matrices will be provided in the context of the upcoming illustrative examples.

The resulting estimates are considered to be quasi-ML. Asymptotically, they approach true ML estimates (Molenaar, 1999; Van Buuren, 1997). The results of such SEM-based approaches for time series models are moment estimates. As first described by Box and Jenkins (1976), recovery of ARMA parameters can be conducted using just the auto-covariances of the time series, and Hamaker and colleagues (Hamaker & Dolan, 2009) demonstrated that this approach results in estimates close to those obtained from SEM software as described here. One benefit of using SEM is the wealth of fit indices. As will be seen, these can be helpful when conducting exploratory analysis.

The following applications demonstrate how to conduct SFA and PFA models within an SEM framework. For these analyses, we use the *lavaan* package (Rosseel, 2012). Code for running these models is available along with the data in the online supplement to this book. We compare model fits of seven CMs: four to ascertain the lag and number of latent variables, and the remaining three to probe the cross-lagged relationships in the PFA.

5.3.1.1 Application 5.1: Exploratory SFA Estimated on Simulated Data with SEM

We can first explore the number of factors and the number of lags by exploring all combinations of $q = (1,2)$ and $l = (1,2)$ for SFA models. For the conditions with $q = 2$, we utilize the known pattern of relations of the observed variables to the latent variables. This is done primarily for simplicity in presentation, and the steps for arriving at the simple structure are identical to those in Chapter 3. The results of the SFA are described here first. We then use this information to build the PFA in the next section and evaluate which model best describes the data.

We begin by describing how to estimate DFM(p,q,2,0,0) which is equivalent to an SFA(2). Please note that all SFA models were generated with this template, albeit with different numbers of latent variables (q) and lags (l) as indicated by the CM (see Table 5.2). The $\mathbf{Y}_{\text{Toep}}$ data matrix of time-embedded variables must contain the p variables up to lag $l = 2$, so the p-observed variables at t-2 will be pasted before the $t - 1$ variables which, in turn, are pasted before the variables at t. Hence, in total, there are 3×6, or $(l+1)p$ variables in the block-Toeplitz (i.e., time-embedded) observed variable matrix.

To describe the matrix structures used to estimate an SFA with SEM, we again depict our matrices with elements "$\lambda_{\#}$" indicating that the element is estimated yet constrained to equal other $\lambda_{\#}$ that have the same subscript, and elements "=1" indicating that the element is fixed at 1. To conduct an exploratory analysis SFA(2) with two latent variables (i.e., DFM(6,2,2,0,0)), the non-zero parameters will be estimated:

$$
\boldsymbol{\Lambda}_{\text{Toep}} = \begin{bmatrix}
=1 & 0 & 0 & 0 & 0 & 0 \\
\lambda_{210} & 0 & 0 & 0 & 0 & 0 \\
\lambda_{310} & 0 & 0 & 0 & 0 & 0 \\
0 & =1 & 0 & 0 & 0 & 0 \\
0 & \lambda_{520} & 0 & 0 & 0 & 0 \\
0 & \lambda_{620} & 0 & 0 & 0 & 0 \\
\lambda_{111} & 0 & =1 & 0 & 0 & 0 \\
\lambda_{211} & 0 & \lambda_{210} & 0 & 0 & 0 \\
\lambda_{311} & 0 & \lambda_{310} & 0 & 0 & 0 \\
0 & \lambda_{421} & 0 & =1 & 0 & 0 \\
0 & \lambda_{521} & 0 & \lambda_{520} & 0 & 0 \\
0 & \lambda_{621} & 0 & \lambda_{620} & 0 & 0 \\
\lambda_{112} & 0 & \lambda_{111} & 0 & =1 & 0 \\
\lambda_{212} & 0 & \lambda_{211} & 0 & \lambda_{210} & 0 \\
\lambda_{312} & 0 & \lambda_{311} & 0 & \lambda_{310} & 0 \\
0 & \lambda_{422} & 0 & \lambda_{421} & 0 & =1 \\
0 & \lambda_{522} & 0 & \lambda_{521} & 0 & \lambda_{520} \\
0 & \lambda_{622} & 0 & \lambda_{621} & 0 & \lambda_{620}
\end{bmatrix} \quad (5.13)
$$

Here, the first two columns contain the loadings of the two shock latent variables at time $t - 2$, $\eta_1(t - 2)$, and $\eta_2(t - 2)$ in $\boldsymbol{\eta}(t - 2)$ on $y_1(t - 2), \ldots, y_6(t - 2)$,

$y_1(t-1), \ldots, y_6(t-1), y_1(t), \ldots, y_6(t)$, as shown in Figure 5.1; the third and fourth columns contain loadings of $\eta_1(t-1)$ and $\eta_2(t-1)$ in $\boldsymbol{\eta}(t-1)$ on the future and contemporaneous observed variables, $y_1(t-1), \ldots, y_6(t-1), y_1(t), \ldots, y_6(t)$; and the final two columns contain the loadings of $\boldsymbol{\eta}(t)$ on the observed variables at t. It can be seen that in this specification of $\boldsymbol{\Lambda}_{Toep}$, latent shock factors at later time points do not influence observed variables at prior time points. Again, the $\boldsymbol{\Phi}_{Toep}$ matrix is constrained to a matrix of zeroes, resulting in a DFM($p,q,l,0,0$). By convention, shock variables are allowed to covary contemporaneously but not at lags: $\text{cov}[\eta(t), \eta(t-u)] = 0$ for all $u > 0$. Thus, $\eta_1(t)$ and $\eta_2(t)$ are allowed to covary with each other, so are $\eta_1(t-1)$ and $\eta_2(t-1)$ as well as $\eta_1(t-2)$ and $\eta_2(t-2)$, but not with the shock variables across lags.

The results of the SFA(2) analysis are as follows:

$$\widehat{\boldsymbol{\Lambda}_{Toep}} = \begin{bmatrix} 1.00^* & 0 & 0 & 0 & 0 & 0 \\ 2.06 & 0 & 0 & 0 & 0 & 0 \\ 1.03 & 0 & 0 & 0 & 0 & 0 \\ 0 & 1.00^* & 0 & 0 & 0 & 0 \\ 0 & 2.00 & 0 & 0 & 0 & 0 \\ 0 & 1.02 & 0 & 0 & 0 & 0 \\ 0.33 & 0 & 1.00^* & 0 & 0 & 0 \\ 0.67 & 0 & 2.06 & 0 & 0 & 0 \\ 0.35 & 0 & 1.03 & 0 & 0 & 0 \\ 0 & 0.53 & 0 & 1.00^* & 0 & 0 \\ 0 & 1.12 & 0 & 2.00 & 0 & 0 \\ 0 & 0.54 & 0 & 1.02 & 0 & 0 \\ 0.00 & 0 & 0.33 & 0 & 1.00^* & 0 \\ 0.14 & 0 & 0.67 & 0 & 2.06 & 0 \\ 0.06 & 0 & 0.35 & 0 & 1.03 & 0 \\ 0 & 0.28 & 0 & 0.53 & 0 & 1.00^* \\ 0 & 0.57 & 0 & 1.12 & 0 & 2.00 \\ 0 & 0.25 & 0 & 0.54 & 0 & 1.02 \end{bmatrix} \tag{5.14}$$

where * indicates that the value was fixed at a non-zero number. We see that the contemporaneous relations are very close to those used to generate the data. For instance, $\hat{\lambda}_{210} = 2.06$ which is very close to the data generating value of 2.00. The lagged relations become subsequently smaller as they become more distant in time. This is a consequence of the weak stationarity exhibited in $\mathbf{y}(t)$, and they will converge to zero over long enough lags as discussed in Chapter 4.

As can be seen in Table 5.2, SFA models with $q = 2$ satisfied all the conventional thresholds for acceptable fit for all fit indices (namely, 1.00 for CFI and TLI, < 0.10 for RMSEA, and < 0.01 for SRMR), whereas those with $q = 1$ did not. The same conclusion was reached using the BIC, which further helped

TABLE 5.2

Fit Indices Resulting from Analysis with Four Candidate SFA Models. DFM(p,q,l,m,n) indicates a dynamic factor model with observed p variables, latent q variables, measurement model lag of l, AR order of m, and MA order n.

Candidate Model (CM):		χ^2	df	CFI	TLI	RMSEA	SRMR	BIC
1.	DFM(6,1,1,0,0)	3088.61	53	0.73	0.66	0.34	0.19	24194
2.	DFM(6,1,2,0,0)	4704.42	133	0.73	0.69	0.17	0.17	36104
3.	DFM(6,2,1,0,0)	84.38	50	1.00	1.00	0.04	0.08	21209
4.	DFM(6,2,2,0,0)	163.23	127	1.00	1.00	0.02	.07	31600

highlight that the SFA model with $q = 2$ and $l = 1$ showed the best fit of all the models considered. Hence, it seems reasonable to conclude that this is a two-factor model of order 1.

5.3.1.2 Application 5.2: PFA Estimated on Simulated Data with SEM

Having arrived at the number of latent variables and lag order, one can either explore the SFA results or continue with the PFA to explore of the AR and cross-lagged relations among the latent variables. Note that this is the data generating model. To estimate a PFA model with SEM, we can set up one CM with matrices as follows:

$$
\boldsymbol{\Lambda}_{\text{Toep}} = \begin{bmatrix} =1 & 0 & 0 & 0 \\ \lambda_{210} & 0 & 0 & 0 \\ \lambda_{310} & 0 & 0 & 0 \\ 0 & =1 & 0 & 0 \\ 0 & \lambda_{520} & 0 & 0 \\ 0 & \lambda_{620} & 0 & 0 \\ 0 & 0 & =1 & 0 \\ 0 & 0 & \lambda_{210} & 0 \\ 0 & 0 & \lambda_{310} & 0 \\ 0 & 0 & 0 & =1 \\ 0 & 0 & 0 & \lambda_{520} \\ 0 & 0 & 0 & \lambda_{620} \end{bmatrix} \quad \boldsymbol{\Psi}_{\text{Toep}} = \begin{bmatrix} \psi_{11} & \psi_{21} & 0 & 0 \\ \psi_{21} & \psi_{22} & 0 & 0 \\ 0 & 0 & \psi_{33} & \psi_{34} \\ 0 & 0 & \psi_{34} & \psi_{44} \end{bmatrix} \qquad (5.15a\text{–}c)
$$

$$
\boldsymbol{\Phi}_{\text{Toep}} = \begin{bmatrix} 0 & 0 & 0 & 0 \\ 0 & 0 & 0 & 0 \\ \phi_{11} & 0 & 0 & 0 \\ \phi_{21} & \phi_{22} & 0 & 0 \end{bmatrix}
$$

with $\boldsymbol{\Theta}_{\text{Toep}}$ being a (12,12)-dimensional diagonal matrix containing the measurement error variances. The lower left-hand (q,q)-dimensioned square of the $\mathbf{\Phi}_{Toep}$ matrix corresponds to the $\mathbf{\Phi}$ matrix in 5.2. Note that the diagonal is freed for estimation, as indicated by the coefficient symbols rather than 0. Those two elements, $\phi_{31_{Toep}}$ and $\phi_{42_{Toep}}$ correspond to ϕ_{11} and ϕ_{22} and represent the AR effects. The estimate for $\phi_{41_{Toep}}$ (ϕ_{21}) is the estimate of the cross-lagged regression coefficient in the regression of $\eta_1(t-1)$ on $\eta_2(t)$. One could test for cross-regression effects by also freeing $\phi_{32_{Toep}}$ to test if the $\eta_2(t-1)$ predicts $\eta_1(t)$, although for now we only fit the known (i.e., simulated) model which we compare to misspecified models next. Additionally, one could free elements in the lower right hand (p,p)-dimensioned portion of the $\mathbf{\Phi}_{Toep}$ matrix, which represents the directed contemporaneous relations among variables at t. In so doing, we arrive at the uSEM model discussed in Chapter 4 (see also Chapter 6).

In the present case, we are estimating a VAR model of the latent variables. As such, $\mathbf{\Psi}$ can be diagonal or may include parameters in the bottom-right portion of the matrix, as this represents the covariance of regression errors at time (t). The lower-right (q,q)-dimensioned sub-matrix contains the covariances of the regression residuals after variance in the latent variables is explained by other variables as indicated via regression of current values on previous values (the coefficients of which are contained in $\mathbf{\Phi}_{Toep}$). The upper-left (q,q)-dimensioned sub-matrix of the $\mathbf{Y}_{Toep}$ matrix contains the covariances of latent variables $t-1$. These covariances represent the contemporaneous relations among variables. Importantly, the submatrix in $\mathbf{Y}_{Toep}$ that relates the variables at $t-1$ with those at t are set to zero. This is consistent with the hypothesis that the cross- and auto-regression parameters in the $\mathbf{\Phi}_{Toep}$ account for these effects.

The lambda values were almost perfectly recovered (standard errors in parentheses): $\hat{\lambda}_{21} = 2.00$ (0.04), $\hat{\lambda}_{31} = 1.02$ (0.03), $\hat{\lambda}_{52} = 1.97$ (0.02), and $\hat{\lambda}_{62} = 0.99$ (0.01). The latent variable covariance and variances, however, evidenced slightly poorer recovery. The residual error variances for were near the true data generating value at $\hat{\psi}_{11} = 2.86(0.31)$ and $\hat{\psi}_{22} = 8.45(0.87)$. The covariance of the residuals from $\eta_1(t)$ and $\eta_2(t)$ also was near the data generating value: $\hat{\psi}_{12} = 2.42(0.40)$. Recall that these values are found in lower right-hand corner of the $\mathbf{\Psi}_{Toep}$ matrix. The data generating $\mathbf{\Phi}_{Toep}$ matrix values were also significant, although the AR coefficient for $\eta_2(t)$ was a bit attenuated while the cross-lag was higher than the data generating model: $\hat{\phi}_{11} = 0.56(0.06)$, $\hat{\phi}_{22} = 0.40(0.06)$, and $\hat{\phi}_{21} = 0.55(0.13)$.

We compared results from this model to two other CMs with the same DFM(6,2,0,1,0) order. They differ only in the presence or absence of cross-lagged effects. CM 5 contains no cross-lagged effects (i.e., neither η_1 nor η_2 predicts the other at a lag); CM 6 contains the wrong effect (i.e., η_1 is predicted by η_2 at a lag), and CM 7 as the true data generation model contains the cross-lagged effect of $\eta_1(t-1)$ on $\eta_2(t)$). We see in Table 5.3 that CM 7 had lower RMSEA, SRMR, and BIC than the other two models. For CM 6, the estimated cross-lagged beta value was not significant ($\psi_{12} = 0.01$, se $= 0.05$, $p = 0.60$), whereas for CM 7 it was ($\psi_{21} = 0.55$, se $= 0.13$, $p < 0.01$). This further supports the selection of CM 7.

TABLE 5.3

Fit Indices Resulting from Analysis with Three Candidate PFA Models

Model:		χ^2	df	CFI	TLI	RMSEA	SRMR	BIC
5.	DFM(6,2,0,1,0)*	87.29	54	1.00	1.00	0.04	0.08	21187
6.	DFM(6,2,0,1,0)**	87.03	53	1.00	1.00	0.04	0.07	21193
7.	DFM(6,2,0,1,0)	48.73	53	1.00	1.00	0.00	0.02	21155

Note: *Model 5 contains no cross-lagged effects among latent variables; **Model 6 contains the wrong cross-lagged effect ($\eta_1(t)$ predicted by $\eta_2(t-1)$). Model 7 is the true data generation model.

5.3.2 SEM with MIIV-2SLS Estimation

MIIV-2SLS estimation arrives at parameter estimates using instrumental variables (IVs). This approach has been found to be more robust than ML estimates when the fitted model may contain misspecifications or under violations of distributional assumptions (Bollen, 1996, 2001), a finding that was replicated in PFA models when compared to the quasi-ML results generated from traditional SEM estimation (Fisher, Bollen, and Gates, 2019). Another benefit of this approach is that because it is a limited-information estimator (as opposed to full-information) and solves each equation separately and it can be used with a far greater number of variables than the previous two approaches.

IVs have long been used to circumvent issues that may arise when an explanatory variable correlates with regression errors. This may occur because of variable omission or errors in the measurement model. A good example of the use of IVs is seen in a study that sought to investigate if there is a relationship between physical attractiveness and income. The problem is that both physical attractiveness and income levels are related to many other variables, such as parental income and health. In this way, they both may be results of common causes. Given this challenge, Glied and Neidell (2010) used a variable that had some shared variance with attractiveness but no shared variance with other confounds as an IV to predict income levels: fluoridation of their country of origin. They reasoned that fluoridation in individuals' town of origin would be an appropriate IV since fluoridation was (1) randomly instituted across the USA with no relationship to country wealth or political affiliations and (2) relates to healthier-looking teeth, which, in turn, relates to attractiveness. By using fluoridation as an IV, the researchers were able to establish a relationship between attractiveness and income that was not confounded with other third sources of shared variance (i.e., parental wealth). Here, fluoridation is an IV (Glied & Neidell, 2010).

Statistically, the issue with using attractiveness to predict income is that the variable "attractiveness" will likely be correlated with the regression error term at the population level, thus causing bias in estimates (Bollen, 1996). This is because, as stated before, both variables could be caused by a large number of other variables, and as such, they have common causes. While the researchers in this example used a clever approach informed by prior research (i.e., there is a relationship between tooth health and attractiveness ratings) to identify an appropriate IV, one could also arrive at IVs by looking at how variables in a given model relate to one another. This model-dependent selection of IVs results in *model-implied instrumental variables* (MIIVs; Bollen, 1996, 2001). Here, based on the model specification itself, a variable is considered an acceptable MIIV for a given equation if it: (A) is correlated with the explanatory variables in the equation and (B) is uncorrelated with the equation error. Two software packages are available for identifying MIIVs, one which is run as a macro within SAS (Bollen & Bauer, 2004) and one that implements the search within R (*MIIVsem*; Fisher et al., 2016). Ideally, an MIIV will also strongly relate to both the predictor and the dependent variable.

A number of generalized methods of moment estimators are available within the MIIV modeling framework. The most common of these is MIIV-2SLS (Two-Stage Least Squares), which will be illustrated here. To use this estimator, the latent variable model is transformed into an observed variable model by selecting one scaling indicator (denoted y_s) per latent variable. This can be done by taking the variable that the researcher believes to best capture the latent construct or it can be empirically derived by examining the degree to which the latent variable explains the variances in the observed variables. Once selected, the lambda values relating to the scaling variables are set to one (the intercept is zero, as is the case throughout our uses in this chapter).

Fisher, Bollen, and Gates (2019) developed a general MIIV-2SLS estimator for DFMs and evaluated its performance in terms of both PFA and SFA model parameters. For pedagogical purposes, we will focus our attention on the PFA model. In the PFA case, the measurement model in Equation (5.1) can be rearranged for these scaling indicators as

$$\boldsymbol{\eta}(t) = \mathbf{y}_s(t) - \boldsymbol{\varepsilon}_s(t). \tag{5.16}$$

From here, we can substitute Equation (5.16) into Equation (5.4), thus removing all latent variables:

$$\begin{aligned} \mathbf{y}_{ns}(t) &= \boldsymbol{\Lambda}(\mathbf{y}_s(t) - \boldsymbol{\varepsilon}_s(t)) + \boldsymbol{\varepsilon}_{ns}(t) \\ \Rightarrow \mathbf{y}_{ns}(t) &= \boldsymbol{\Lambda}\mathbf{y}_s(t) + \left[\boldsymbol{\varepsilon}_{ns}(t) - \Lambda\boldsymbol{\varepsilon}_s(t)\right] \end{aligned} \tag{5.17}$$

where the composite error terms are in the brackets of the second equation. The subscript "s" in these equations indicates terms related to the scaling indicator and the subscript "ns" the non-scaling indicators.

The latent variable model ("structural" model in the state space sense) can be similarly revised with the observed variable transformation:

$$\mathbf{y}_s(t) = \boldsymbol{\phi}_1\left[\mathbf{y}_s(t-1) + \boldsymbol{\varepsilon}_s(t-1)\right] + \ldots + \boldsymbol{\phi}_m\left[\mathbf{y}_s(t-m) + \boldsymbol{\varepsilon}_s(t-m)\right] + \boldsymbol{\zeta}(t) \tag{5.18a}$$

For clarity, the composite errors can be gathered as follows in the brackets:

$$\mathbf{y}_s(t) = \sum_{u=1}^{m} \boldsymbol{\phi}_u \mathbf{y}_s(t-u) + \left[\sum_{u=1}^{m} \boldsymbol{\phi}_u \boldsymbol{\varepsilon}_s(t-u) + \boldsymbol{\zeta}(t) \right]. \tag{5.18b}$$

OLS is not appropriate since the composite error term will be correlated with the dependent variable. The MIIVs will be referred to here as **V**, an $T \times (O_v + 1)$ matrix where the first column is a vector of ones and the T rows represent observations of the O_v MIIVs. The matrix **Z** is a block-diagonal matrix containing the latent variables that have a direct effect on y, and $\boldsymbol{\Gamma}$ contains parameter estimates. In the two-stage estimation procedure, these are predicted using the MIIVs

$$\hat{\mathbf{Z}} = \mathbf{V}(\mathbf{V}'\mathbf{V})^{-1}\mathbf{V}'\mathbf{Z} \tag{5.19}$$

which are then used to arrive at the parameters contained in $\boldsymbol{\Gamma}_{IV}$:

$$\widehat{\boldsymbol{\Gamma}_{\text{IV}}} = \left(\hat{\mathbf{Z}}'\hat{\mathbf{Z}}\right)^{-1}\hat{\mathbf{Z}}'\mathbf{y}. \tag{5.20}$$

As with estimation using SEM explained in Section 5.2.1, one first must time-embed the data by copying variables and removing lines to generate lags. The "Toep" matrices described under the SEM estimation approach apply here as well. A convenient outcome of conducting MIIV-2SLS in the context of time series data is that the lag-1 non-scaling observed variables will always be appropriate IVs for equations pertaining to the lag-0 estimates, such as measurement model and relations among lagged and contemporaneous latent variables. This reduces the search space for appropriate MIIVs.

5.3.2.1 Application 5.3: PFA Estimated on Simulated Data with MIIV-2SLS

To estimate the data example with MIIV-2SLS, we utilize the R package *MIIVsem* (Fisher et al., 2016). Model specification can occur using similar syntax as in the *lavaan* approach used above but the *miive* function in the *MIIVsem* package. Results were very similar to those seen for the quasi-ML approach. Again, the λ values nearly matched the data-generating values, and the AR coefficient for $\eta_2(t)$ was a bit attenuated while the cross-lag was higher than the data-generating model. The covariance/variance matrix of the latent variables cannot be estimated via this approach, and as such fit indices are unavailable. While this is a limitation of the MIIV-2SLS approach, a major benefit of this approach is that estimates are more reliable than quasi-ML estimates with traditional SEM packages, particularly when model misspecifications are present given the robustness of the MIIV-2SLS estimator. In fact, model misspecifications occurring at the structural level do not impact the estimates of the measurement model at all, and in many cases, misspecifications in the structural and measurement model do not impact estimates for other equations (Bollen, Gates, & Fisher, 2018). The online supplement provides examples of correctly and misspecified models.

5.3.2.2 Application 5.4: PFA on fMRI Data

We demonstrate the ability to arrive at estimates for a large number of variables by conducting a confirmatory PFA on fMRI data collected by Dr. Kristen Lindquist (described in Chapter 1). The data used here are 75 observed variables (i.e., brain regions of interest) of length $T = 450$ that load onto 10 latent constructs (termed, "brain networks", in fMRI literature) see Figure 5.2. The dashed lines among the networks represent lagged relations. For instance, as suggested by the negative sign, the language network suppresses the dorsal default mode network (dDMN) at the next time point. This follows previous literature since the dDMN is a network that surfaces when the brain is at rest. As such, it is often negatively correlated with networks that indicate focused tasks. The loadings of how the brain regions relate to the latent network quantify how integral the latent network activity is to respective brain regions' activity.

5.3.3 Raw Data Likelihood Approach

The Kalman filter (Kalman, 1960; Kalman & Bucy, 1961) was initially developed for generating minimum mean square error forecasts for the guidance, navigation, and control of vehicles and related engineering applications. However, since its inception, the estimation approach has been exploited for model estimation purposes in many applications in physics, economics, biology, statistics, and other physical as well as social and behavioral sciences. As will be described below, the Kalman filter provides a way to obtain recursive (to be defined next) least-squares estimates of the latent factor scores and in the process, provides by-products that can be used to compute the raw data likelihood of any state-space model.

To begin, we re-express the DFM(p,q,l,m,n) in an alternative form, known as the state-space form. We first note that multiple elements in the DFM(p,q,l,m,n) are latent (unobserved) variables, including: $\boldsymbol{\eta}(t-u)$, with $u = 0,\ldots l$, and $u = 0,\ldots m$ in Equations (5.1) and (5.2), respectively; $\boldsymbol{\varepsilon}(t)$; and $\boldsymbol{\zeta}(t-u)$, with $u = 0,\ldots,n$. Furthermore, except for the measurement errors in $\boldsymbol{\varepsilon}(t)$, all of the latent variables appear in concurrent (at time t) and lagged (at time $t-u$) forms. To implement the transformation into state-space form, the caveat here is to put latent variables whose lagged versions are needed into an expanded latent variable series, denoted herein as $\boldsymbol{\eta}_{SS}(t)$. This is conceptually similar to the approach used in SEM in that additional lagged variables are appended to the data set in order to obtain estimates. A primary difference here is that included in this expanded latent variable series are lagged versions of both $\boldsymbol{\eta}(t)$ and $\boldsymbol{\zeta}(t)$ up to the desired orders: namely, $\max(l, m-1)$, whichever one is larger; and n, respectively. In this way, the analysis enables MA or VARMA approaches.

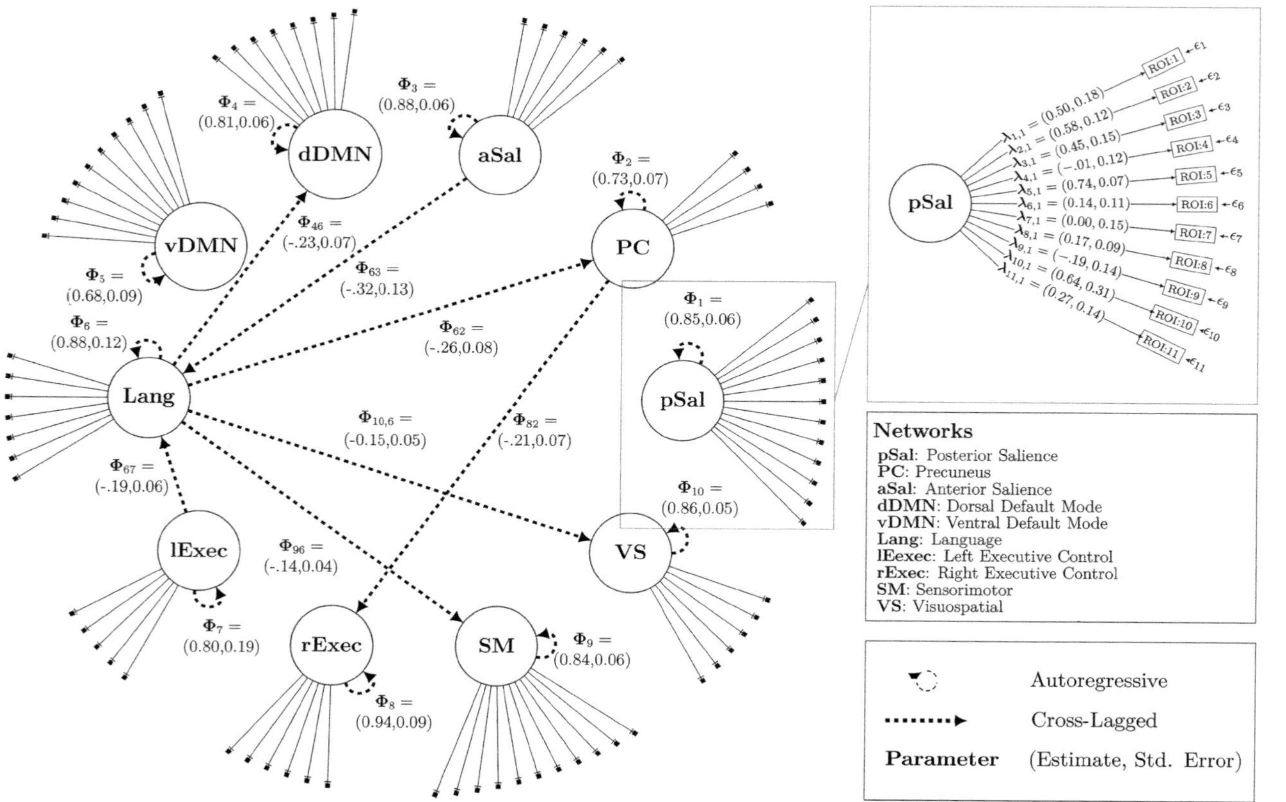

FIGURE 5.2
Example of PFA conducted on fMRI data and estimated with MIIV-2SLS.

Thus, we can rewrite Equation (5.1) in an expanded form as

$$\mathbf{y}_{SS} = \mathbf{\Lambda}_{SS}\boldsymbol{\eta}_{SS}(t) + \boldsymbol{\varepsilon}(t), \tag{5.21}$$

where the expanded latent variable vector, $\varsigma_{KF}(t)$, consists of:

$$\boldsymbol{\eta}_{SS}(t) = \begin{bmatrix} \boldsymbol{\eta}(t) \\ \boldsymbol{\eta}(t-1) \\ \vdots \\ \boldsymbol{\eta}(t-\max(l-1, m-2)) \\ \boldsymbol{\eta}(t-\max(l, m-1)) \\ \boldsymbol{\zeta}(t) \\ \boldsymbol{\zeta}(t-1) \\ \vdots \\ \boldsymbol{\zeta}(t-n-1) \\ \boldsymbol{\zeta}(t-n) \end{bmatrix}, \tag{5.22}$$

and,

$$\mathbf{\Lambda}_{SS} = \begin{bmatrix} \overbrace{\mathbf{\Lambda}(0)}^{p\times q} & \overbrace{\mathbf{\Lambda}(1)}^{p\times q} & \cdots & \overbrace{\mathbf{\Lambda}(l)}^{p\times q} & \mathbf{0}_{p\times(q*\max(0,m-l-1))} & \mathbf{0}_{p\times q(n+1)} \end{bmatrix}, \tag{5.23}$$

in which $\mathbf{0}_{r\times s}$ denotes an $r \times s$ matrix of zeros, and $\mathbf{0}_{p\times(q^*\max(0,m-l-1))}$ is simply a matrix of "filler" zeros if the AR order m is larger than the shock order l. The covariance matrix of measurement errors, $\mathbf{\Theta}$, is similarly expanded to include zeros.

In a similar vein, $\mathbf{\Phi}$ in Equation (5.2) also has to be adapted to reflect the relations among the (same set of) manifest variables and the set of expanded latent variables $\boldsymbol{\eta}_{SS}(t)$. The expanded form of $\mathbf{\Phi}$, denoted below as $\mathbf{\Phi}_{SS}$:

$$\mathbf{\Phi}_{SS} = \begin{bmatrix} \mathbf{\Phi}(1) & \mathbf{\Phi}(2) & \cdots & \mathbf{\Phi}(\max(l, m-1)) & \mathbf{\Phi}(\max(l+1, m)) & \mathbf{0}_{q\times q} & \boldsymbol{\theta}(1) & \cdots & \boldsymbol{\theta}(n-1) & \boldsymbol{\theta}(n) \\ \mathrm{I}_q & 0_{q\times q} & \cdots & 0_{q\times q} & 0_{q\times q} & 0_{q\times q} & 0_{q\times q} & \cdots & 0_{q\times q} & 0_{q\times q} \\ 0_{q\times q} & \ddots & \ddots & \vdots & \vdots & \vdots & \vdots & \vdots & & \vdots \\ 0_{q\times q} & \cdots & \ddots & 0_{q\times q} & 0_{q\times q} & 0_{q\times q} & 0_{q\times q} & \cdots & 0_{q\times q} & 0_{q\times q} \\ 0_{q\times q} & \cdots & \cdots & \mathrm{I}_q & 0_{q\times q} & 0_{q\times q} & 0_{q\times q} & \cdots & 0_{q\times q} & 0_{q\times q} \\ 0_{q\times q} & \cdots & \cdots & \cdots & \cdots & \cdots & \cdots & \cdots & 0_{q\times q} & 0_{q\times q} \\ 0_{q\times q} & \cdots & \cdots & \cdots & \cdots & \mathrm{I}_q & 0_{q\times q} & \cdots & 0_{q\times q} & 0_{q\times q} \\ \vdots & \vdots & \vdots & \vdots & \vdots & \vdots & \ddots & \vdots & \vdots & \vdots \\ 0_{q\times q} & \cdots & \cdots & \cdots & \cdots & \cdots & \cdots & \ddots & 0_{q\times q} & 0_{q\times q} \\ 0_{q\times q} & \cdots & \cdots & \cdots & \cdots & 0_{q\times q} & 0_{q\times q} & \cdots & \mathrm{I}_q & 0_{q\times q} \end{bmatrix} \tag{5.24}$$

where I_q denotes a $q \times q$ identity matrix; $\boldsymbol{\Phi}(u)$, as defined in the context of Equation (5.2), is a $q \times q$ matrix of auto- and cross-regression coefficients linking the latent variables from u time points ago (i.e., at lag u) to the latent variables at time t. If $m < l$, then $\boldsymbol{\Phi}(m+1)\ldots\boldsymbol{\Phi}(l)$ are $q \times q$ matrices of zeros. We implicitly assume in the specifications above that if the values of l and m are such that the value u enclosed in the parentheses in $\boldsymbol{\Phi}(u)$ is equal to zero, then the entire column block of $\boldsymbol{\Phi}_{SS}$ containing that entry may be omitted together with the corresponding latent variables in $\boldsymbol{\eta}_{SS}(t)$. Finally, we define a linkage matrix, $\mathbf{Z}$ (a different matrix than used in MIIV-2SLS estimation), that serves to relate elements in $\boldsymbol{\eta}_{SS}(t)$ to the process noises in $\boldsymbol{\zeta}(t)$ as

$$\mathbf{Z} = \begin{bmatrix} \mathrm{I}_q \\ 0_{q\times q} \\ \vdots \\ 0_{q\times q} \\ 0_{q\times q} \\ \mathrm{I}_q \\ 0_{q\times q} \\ \vdots \\ 0_{q\times q} \\ 0_{q\times q} \end{bmatrix} \tag{5.25}$$

One can then obtain an expanded process noise vector, $\boldsymbol{\zeta}_{SS}(t) = \mathbf{Z}\boldsymbol{\zeta}(t)$. With these expanded elements, the DFM$(p,q,1,m,0)$ model can now be rewritten in the so-called state-space form as

$$\boldsymbol{\eta}_{SS}(t) = \boldsymbol{\Phi}_{SS}\ \boldsymbol{\eta}_{SS}(t-1) + \boldsymbol{\zeta}_{SS}(t). \tag{5.26}$$

The linear state-space model, in the absence of intercept terms, coincides in form with the PFA model shown in Equations 5.3 and 5.4. For instance, a PFA(1,0) model can be formulated as a state-space model with:

$$\boldsymbol{\eta}_{SS}(t) = \left[\boldsymbol{\eta}(t)\right] \tag{5.27a}$$

$$\boldsymbol{\Phi}_{SS} = \left[\boldsymbol{\Phi}(1)\right] \tag{5.27b}$$

$$\boldsymbol{\zeta}_{SS} = \left[\boldsymbol{\zeta}(0)\right] \tag{5.27c}$$

With appropriate reorganization of elements that do show serial dependencies over time into an expanded vector of latent variables, $\boldsymbol{\eta}_{SS}(t)$, the SFA model shown in Equation (5.1) can also be structured as a special case of the state-space model. An SFA(l) model can be formulated as a state-space model with:

$$\boldsymbol{\eta}_{SS}(t) = \begin{bmatrix} \boldsymbol{\eta}_i(t) \\ \boldsymbol{\eta}_i(t-1) \end{bmatrix} \tag{5.28a}$$

$$\Phi_{ss} = \begin{bmatrix} \mathbf{0}_{q\times q} \\ \mathbf{0}_{q\times q} \end{bmatrix} \tag{5.28b}$$

$$\Lambda_{SS} = \begin{bmatrix} \Lambda(0) \\ \Lambda(1) \end{bmatrix} \tag{5.28c}$$

Stationarity assumptions, as discussed earlier in Chapter 4, require that all patterns of intra-individual variability are functions of lag order and not functions of time. That is, the assumption of stationarity is that parameter estimates are *time-invariant*. However, in other applications involving alternative models of change, the state-space model of the form in Equations 5.1 and 5.2 can actually be used to model selected kinds of non-stationary change processes, as we will illustrate in Chapter 7.

The Kalman filter comprises two sets of equations: *prediction* and *updating* (or *correcting*) equations. The *initial conditions* of the latent variables are defined as $\boldsymbol{\eta}_{SS}\ (0|0){:} = \boldsymbol{\mu}_0$, $\boldsymbol{\Sigma}_{\eta SS}(0|0){:} = \boldsymbol{\Sigma}_{\eta SS,0}$, where $\boldsymbol{\Sigma}_{\eta SS,0}$ is the covariance matrix of the latent variables prior to the first available time point. With these initial conditions set, the *prediction step* comprises the following for $(1 \le t \le T)$:

$$\begin{aligned}
&\eta_{SS}(t\,|\,t-1) = \Phi_{SS}\eta_{SS}(t-1\,|\,t-1),\\
&\Sigma_{\eta SS}(t\,|\,t-1) = \Phi_{SS}\,\Sigma_{\eta}(t-1\,|\,t-1)\Phi_{SS}{}' + \Psi_{SS},\\
&\mathbf{y}(t\,|\,t-1) = \Lambda_{SS}\eta_{SS}(t\,|\,t-1),\\
&\Sigma_{ySS}(t\,|\,t-1) = \Lambda_{SS}\eta_{SS}(t\,|\,t-1)\Lambda_{SS}{}' + \Xi_{SS}.
\end{aligned} \tag{5.29a–d}$$

Here, the Λ_{SS} measurement matrices are assumed to be known and non-stochastic. $\boldsymbol{\Sigma}_{\eta SS}\ (t-1|t-1)$ provides the variance-covariance matrix of $\boldsymbol{\eta}_{SS}$ at $t-1$, given $t-1$, Ψ_{SS} **is** the $q\ x\ q$ variance-covariance matrix of the regression errors, and $\Phi_{SS}\ p\ x\ p$ is the variance-covariance matrix of the measurement errors. The *correction step* is

$$\eta_{SS}(t\,|\,t) = \eta_{SS}(t\,|\,t-1) + \mathbf{K}(t)\mathbf{v}(t)) \tag{5.30a}$$

$$\Sigma_{\eta SS}(t\,|\,t) = [\mathbf{I}_q - \mathbf{K}(t)\Lambda_{SS}]\Sigma_{\eta}(t\,|\,t-1), \tag{5.30b}$$

where

$$\mathbf{K}(t) := \Sigma_{\eta}(t\,|\,t-1)\Lambda_{SS}{}'\left[\Lambda_{SS}\,\Sigma_{\eta}(t\,|\,t-1)\Lambda_{SS}{}' + \Theta_{SS}\right]^{-1}. \tag{5.30c}$$

$\mathbf{K}(t)$ is the *Kalman filter gain* matrix which is multiplied by a vector of prediction errors $\mathbf{v}(t) = y(t)\text{-}y(t|t-1)$. I_q a $q \times q$ identity matrix. The terms are byproducts of the KF that can be used for parameter estimation purposes, and are useful for diagnostic purposes, as will be shown later.

This process is described as recursive as it involves sequential factor score estimation one time point at a time, beginning with the prediction step for $t = 1$, followed by the correction step for $t = 1, \ldots, T$. Hence, first $\boldsymbol{\Sigma}_{\eta}(1|0)$ is computed according to the second equation in 5.30: $\boldsymbol{\Sigma}_{\eta SS}(1|0) = \boldsymbol{\Phi}_{SS}\, \boldsymbol{\Sigma}_{\eta}(0|0)\, \boldsymbol{\Phi}_{SS}' + \boldsymbol{\Psi}_{SS}$. Next, the correction step is implemented to arrive at $\boldsymbol{\Sigma}_{\eta SS}(1|1)$ and $\boldsymbol{\eta}_{SS}(t|t)$. These steps are then repeated for $t = 2, 3, \ldots, T$, by iteratively inserting the updated values as time progresses. The estimates for $\boldsymbol{\eta}_{SS}(t|t)$ thus contain two parts: the predicted value based on information available at $t - 1$ and the correction based on new observed information entering at time t.

As noted in the introduction of this section, the KF was initially conceived for the purposes of real-time navigation purposes as data arrive in real time. As such, *filtering* provides online (real-time) estimates of the latent variables at time t based on information from time $t = 1, \ldots, t$. Forecasting occurs when one performs the estimation of the latent variables at time t which occurs after T based on all information available up to T (e.g., using the prediction step without the correction step at time t). Almost all of the applications covered in this book are offline estimation, namely, estimation is performed after all the data have arrived or have been collected. Nonetheless, we provide the *forecasting step* for $t > T$:

$$\mathbf{H}_{ss}(T+t\,|\,T) = \boldsymbol{\Phi}_{ss}\boldsymbol{\eta}_{ss}(T+t-1\,|\,T), \tag{5.31a}$$

$$\Sigma_{\eta ss}(T+t\,|\,T) = \boldsymbol{\Phi}_{ss}\,\Sigma_{\eta ss}(T+t-1\,|\,T)\boldsymbol{\Phi}_{ss}{}' + \boldsymbol{\Psi}_{ss}, \tag{5.31b}$$

$$\mathbf{y}(T+t\,|\,T) = \boldsymbol{\Lambda}_{ss}\boldsymbol{\eta}_{ss}(T+t\,|\,T), \tag{5.31c}$$

$$\Sigma_{\eta ss}(T+t\,|\,T) = \boldsymbol{\Lambda}_{ss}\boldsymbol{\eta}_{ss}(T+t\,|\,T)\boldsymbol{\Lambda}_{ss}{}' + \Theta_{ss}. \tag{5.31d}$$

Once all data are obtained and prediction and correcting conducted, researchers often using *smoothing* to further refine estimates. We describe here one kind of smoothing procedure, the fixed interval smoother (Anderson & Moore, 1979; Bar-Shalom, Li, & Kirubarajan, 2001; Dolan & Molenaar, 1991), which performs latent variable estimation based on all observed information up to time T. Smoothing is essentially a process for performing latent variable estimation for each time $t < T$ using observed information up to and beyond time t. The fixed interval smoother, in particular, can be implemented by first performing filtering forward in time from $t = 1, \ldots T$, followed by refinement of the latent variable estimates backward in time for time $t = T, \ldots, 1$. Starting at time T with $\boldsymbol{\eta}_{SS}(t|T)$

and $\boldsymbol{\Sigma}_{\eta ss}(t|T)$ from filtering, the fixed interval smoother is performed for $t = T - 1, \ldots, 1$ as

$$\boldsymbol{\eta}_{ss}(t\,|\,T) = \boldsymbol{\eta}_{ss}(t\,|\,t) + \mathbf{J}(t)(\boldsymbol{\eta}_{ss}(t+1\,|\,T) - \boldsymbol{\eta}_{ss}(t+1\,|\,t)) \tag{5.32a}$$

$$\Sigma_{\eta_{ss}}(t\,|\,T) = \Sigma_{\eta_{ss}}(t\,|\,t) + \mathbf{J}(t)\left[\Sigma_{\eta_{ss}}(t+1\,|\,T) - \Sigma_{\eta_{ss}}(t+1\,|\,t)\right]\mathbf{J}(t), \tag{5.32b}$$

where $\mathbf{J}(t) = \boldsymbol{\Sigma}_{\eta ss}(t|t)\, \boldsymbol{\Phi}_{ss}'\, \boldsymbol{\Sigma}^{-1}_{\eta ss}(t+1|t)$.

Raw data ML estimation is then carried out to obtain estimates of all parameters in the vector, δ. The log-likelihood function for parameter estimation purposes can be computed readily using by-products of the KF as

$$\ln l(\delta) = -\frac{pT}{2}\ln 2\pi - \frac{1}{2}\sum_{t=1}^{T}[\ln|\Sigma_y^{-1}(t\,|\,t-1)| - \boldsymbol{\varepsilon}(t)'\Sigma_y(t|\,t-1)\,\boldsymbol{\varepsilon}(t)] \tag{5.33}$$

where $\boldsymbol{\varepsilon}(t) := \mathrm{y}(t) - \mathbf{y}(t\,|\,t-1)$ and Σ_y is the associated covariance matrix for one-step-ahead prediction errors. The log-likelihood function is sometimes referred to as the prediction error decomposition function (Ljung & Caines, 1979; Schweppe, 1965).

The relative costs and benefits of utilizing Kalman filter approaches over SEM-based methods have received much attention. In terms of bias in the estimates, some have found that the Kalman filter approach outperforms the quasi-ML estimation seen in SEM above, especially in yielding accurate standard error estimates (e.g., Chow et al., 2010; Voelkle et al., 2012; Zhang, Hamaker, & Nesselroade, 2008). Others have reported that in some cases the biases of the point estimates obtained via SEM are quite similar to those found in raw data likelihood approaches (Hamaker, Dolan, & Molenaar, 2002, 2003). A primary benefit of KF is that it is amenable to the estimation of time-varying effects, which is discussed in Chapter 7.

5.3.3.1 Application 5.4: PFA Estimated on Simulated Data with the Kalman Filter

Here, we utilize an R package, Dynamic Modeling in R (*dynr*; Ou, Hunter, & Chow, 2017). *dynr* implements the fixed interval smoother and parameter estimation via the prediction error decomposition function. The R package *dynr* also supports the estimation of linear and non-linear discrete-time state-space models and linear and non-linear continuous-time models, which will be covered later in Chapter 7. The input files providing the code are available in the online supplement for this book. We take as a given that the number of lags is one and the factor structure is identical to what was found in the previous set of analyses and go straight to the PFA model with a lag of one and two latent variables. Results (see supplemental code) indicate point estimates as being similar to the SEM approaches.

5.4 Conclusions

In this chapter, we presented a comprehensive introduction to DFMs. By limiting our presentation mainly to the two most common specifications of this model—PFA and SFA—we focused on approaches that are likely to be most useful for readers. However, please remember that any number of constraints are possible so long as the model is identified. Additional specifications are covered in the next chapter, where we continue with a discussion of obtaining data-driven models.

Three estimation procedures were presented here: SEM conducted with quasi-ML, SEM conducted with MIIV-2SLS, and KF. Our toy example here did not indicate great differences among these methods in terms of point estimates or standard errors. Research is ongoing regarding the relative benefits of these estimation approaches. With continued advancements in this area, we suggest the reader become aware of current views and evidence regarding the benefits of each procedure. Our account of the estimation procedures served to provide an introduction to estimation in this context, bearing in mind that research in this area is ongoing.

References

Anderson, B.D.O. & Moore, J.B. (1979). *Optimal filtering.* Englewood Cliffs, NJ: Prentice Hall.

Anderson, T.W. (1963). The use of factor analysis in the statistical analysis of multiple time series. *Psychometrika, 28*(1), 1–25.

Bai, J. & Wang, P. (2012). *Identification and estimation of dynamic factor models.* Germany: University Library of Munich.

Bai, J.& Wang, P. (2014). Identification theory for high dimensional static and dynamic factor models. *Journal of Econometrics, 178*(2), 794–804.

Bar-Shalom, Y., Li, X.R., & Kirubarajan, T. (2001). *Estimation with applications to tracking and navigation: Theory algorithms and software.* New York, NY: Wiley.

Bollen, K.A. (1996). An alternative two stage least squares (2SLS) estimator for latent variable equations. *Psychometrika, 61*(1), 109–121.

Bollen, K.A. (2001). Two-stage least squares and latent variable models: Simultaneous estimation and robustness to misspecifications. *Structural Equation Modeling: Present and Future,* 119–138.

Bollen, K.A. & Bauer, D.J. (2004). Automating the selection of model-implied instrumental variables. *Sociological Methods & Research, 32*(4), 425–452.

Bollen, K.A., Gates, K.M., & Fisher, Z. (2018). Robustness conditions for MIIV-2SLS when the latent variable or measurement model is structurally misspecified. *Structural Equation Modeling: A Multidisciplinary Journal, 25*(6), 848–859.

Box, G.E. & Jenkins, G.M. (1976). *Time series analysis: Forecasting and control.* San Francisco, CA: Holden-Day.

Browne, M.W. & Nesselroade, J.R. (2005). Representing psychological processes with dynamic factor models: Some promising uses and extensions of ARMA time series models. In A. Maydeu-Olivares & J.J. McArdle (Eds.), *Psychometrics: A festschrift to Roderick P. McDonald* (pp. 415–452). Mahwah, NJ: Lawrence Erlbaum Associates Publishers.

Browne, M.W. & Zhang, G. (2007). Developments in the factor analysis of individual time series. In R. Cudeck & R.C. MacCallum (Eds.), *Factor analysis at 100: Historical developments and future directions*. Mahwah, NJ: Erlbaum.

Chow, S.M., Ho, M.H.R., Hamaker, E.L., & Dolan, C.V. (2010). Equivalence and differences between structural equation modeling and state-space modeling techniques. *Structural Equation Modeling, 17*(2), 303–332.

Dolan, C.V. & Molenaar, P.C.M. (1991). A note on the calculation of latent trajectories in the quasi Markov simplex model by means of regression method and the discrete Kalman filter. *Kwantitatieve Methoden, 38*, 29–44.

Ferrer, E. & Nesselroade, J.R. (2003). Modeling affective processes in dyadic relations via dynamic factor analysis. *Emotion, 3*(4), 344.

Fisher, Z.F., Bollen, K.A., & Gates, K.M. (2019). A limited information estimator for dynamic factor models. *Multivariate Behavioral Research, 54*(2), 246–263.

Fisher, Z.F., Bollen, K.A., Gates, K.M., & Rönkkö, M. (2016). Package 'MIIVsem'. [Computer software]. https://cran.rstudio.com/web/packages/MIIVsem/index.html

Glied, S. & Neidell, M. (2010). The economic value of teeth. *Journal of Human Resources, 45*(2), 468–496.

Hamaker, E.L. & Dolan, C.V. (2009). Idiographic data analysis: Quantitative methods—From simple to advanced. In J. Valsiner, P. Molenaar, M. Lyra, & N. Chaudhary (Eds.), *Dynamic process methodology in the social and developmental sciences* (pp. 191–216). New York, NY: Springer.

Hamaker, E.L., Dolan, C.V., & Molenaar, P.C. (2002). On the nature of SEM estimates of ARMA parameters. *Structural Equation Modeling, 9*(3), 347–368.

Hamaker, E.L., Dolan, C.V., & Molenaar, P.C. (2003). ARMA-based SEM when the number of time points T exceeds the number of cases N: Raw data maximum likelihood. *Structural Equation Modeling, 10*(3), 352–379.

Kalman, R.E. (1960). A new approach to linear filtering and prediction problems. *Journal of Basic Engineering, 82*(1), 35–45.

Kalman, R.E. & Bucy, R.S. (1961). New results in linear filtering and prediction theory. *Journal of Basic Engineering, 83*(3), 95–108.

Ljung, L. & Caines, P.E. (1979). Asymptotic normality of prediction error estimators for approximate system models. *Stochastics, 3*(1), 26–49.

Molenaar, P.C. (1985). A dynamic factor model for the analysis of multivariate time series. *Psychometrika, 50*(2), 181–202.

Molenaar, P.C. (2017). Equivalent dynamic models. *Multivariate Behavioral Research, 52*(2), 242–258.

Nesselroade, J.R., McArdle, J.J., Aggen, S.H., & Meyers, J.M. (2002). Dynamic factor analysis models for representing process in multivariate time-series. In D.S. Moskowitz & S.L. Hershberger (Eds.), *Modeling intraindividual variability with repeated measures data: Methods and applications* (pp. 235–265). Mahwah, NJ: Lawrence Erlbaum Associates Publishers.

Ou, L., Hunter, M. D., & Chow, S. M. (2017). What's for dynr: A package for linear and nonlinear dynamic modeling in R. *Journal of Statistical Software*, 11(1), 91–111.

Rosseel, Y. (2012). lavaan: An R package for structural equation modeling. *Journal of Statistical Software, 48,* 1–36.

Schweppe, F.C. (1965). Evaluation of likelihood functions for Gaussian signals. *IEEE Transactions on Information Theory, IT-11,* 11(1), 61–70.

Van Buuren, S. (1997). Optimal transformations for categorical autoregressive time series. *Statistica Neerlandica, 51*(1), 90–106.

Voelkle, M.C., Oud, J.H., Davidov, E., & Schmidt, P. (2012). An SEM approach to continuous time modeling of panel data: Relating authoritarianism and anomia. *Psychological Methods, 17*(2), 176.

Zhang, Z., Hamaker, E.L., & Nesselroade, J.R. (2008). Comparisons of four methods for estimating a dynamic factor model. *Structural Equation Modeling, 15*(3), 377–402.

6

Model Specification and Selection Procedures

When studying the individual, *a priori* or confirmatory models of how variables relate to each other across time may not be known. Exploratory P-technique (Chapter 3) directly attends to this by finding the relations between latent constructs and the observed variables with minimal input provided by the researcher. Similarly, identifying the lag order in AR and VAR analyses (as seen in Chapter 4) represents a data-driven approach for model selection. In Chapter 5, we examined model fit indices as well as significance tests to decide which paths among latent variables to include in the model. Here, we extend this line of best practices by formally describing various methods for arriving at patterns of relations among observed or latent variables that are computationally efficient and easy for researchers to implement.

By "patterns of relations" we mean the structural form of the model rather than actual numeric estimates. The focus of model building is to obtain a model that describes *how* and *if* variables relate across time by detecting which relations to include in analysis. The resulting patterns of relations can be considered a qualitative description of the process, where the numeric estimates for parameters associated with any relations are the quantitative aspects of the process. With data-driven searches, researchers can examine the individual-level patterns to arrive at a better understanding of systematic differences between individuals. As an example, one might ask if individuals with a relation between daily stress and alcohol use tend to drink more heavily in general. Another example is seen in the descriptive interpretation of the pattern of relations, which can be particularly illuminating for person-centered clinical practice (as discussed by Fisher & Boswell, 2016). For instance, consider an individual whose experience of pain relates to irritability. In this case, one might want to test if an intervention addressing pain management issues aids in resolving downstream interpersonal issues. For another individual, perhaps this relation does not exist and pain is not related to, or central to, irritability. This would have implications for potential treatment and intervention efforts to attempt. Experimentation would need to be conducted to ascertain causality.

Another reason for data-driven searches is that hypotheses may not exist for how constructs relate across time for individuals. It has long been known that those with depression-related diagnoses also tend to have anxiety; however, it is not clear how this manifests across time for individuals. Does anxiety tend to occur before depression, or the reverse, or do they tend to co-occur

DOI: 10.1201/9780429172649-6

for a given individual? Data-driven approaches can help to clarify these types of issues at this nascent stage of discovery. Finally, allowing for data-driven results enables the field to identify what dynamic patterns might be considered typical or generalizable. From this understanding, one can begin to generate hypotheses regarding why some individuals may differ from typical dynamic patterns. This chapter focuses specifically on model specification and selection to arrive at person-specific patterns of relations by exploring the data. It might help the reader to keep in mind the potential research questions of interest that can be addressed from these data-driven searches.

Model specification procedures encompass a range of approaches for modifying the structural form of models. Here, we present a foundational overview of model specification that covers a wider span of model specification approaches than those previously discussed. Model specification can be conducted on a spectrum ranging from entirely data-driven, exploratory methods to confirmatory models, where only slight, if any, modifications are considered. Most of the procedures introduced below can be used along this spectrum.

The chapter is organized as follows. We first discuss general methods for modifying individual models that can be used within many modeling approaches. The objective of these approaches is to identify which features, or relations among variables, to include when estimating a final model. From this perspective, we introduce general classes of feature selection that can be used for model specification and modification: filter, wrapper, and embedded methods. We focus model building from within a time series perspective in our examples. Given that individual-level model searches are often influenced by noise (Smith et al., 2011), we end this chapter with a description of approaches that allow for individual-level nuances while also capitalizing on shared information across individuals to help detect signal from noise.

6.1 Data-Driven Methods for Person-Specific Discovery of Relations among Variables

In what follows, we describe machine learning approaches, which can broadly be defined as any method that extracts patterns from data (Goodfellow, Bengio, & Courville, 2016). The primary goal of the methods considered here is to select the relations among variables that best describe the dynamic qualities of the process that gave rise to the data. The challenge is to recover the true relations while not over-fitting to the data or modeling noise. For this reason, most approaches favor parsimony (Mumford & Ramsey, 2014). This is particularly salient in the study of individuals, as individual-level nuances might emerge, and it is difficult to know for certain when they truly exist for that individual or are simply caused by noise or by the omission of other potential predictors in the search space. As such, all results must be interpreted with caution, and considered potentially hypothesis-generating. When the goal is to make between-person inferences, the ensuing intensive longitudinal models require validation with external data (Aldenderfer & Blashfield, 1984), such as a static characteristic of the individual (e.g., those with a specific pattern of relations tend to have a specific diagnosis).

Modification procedures tend to perform best when the model structure is already close to the true model (e.g., MacCallum, Roznowski, & Necowitz, 1992). Hence, it is recommended that exploratory approaches utilize any information available *a priori* to start the model search at a point that may be closer to the individual's "true" model. This may be influenced by prior knowledge or be data-driven. After introducing fundamental concepts, we highlight this issue and offer a description of approaches that improve upon individual-level searches. In short, these approaches first identify relations that replicate across individuals. Starting the exploratory search for individuals with relations that have been found to be consistent across the sample greatly improves the recovery rates of true model features, and we end with examples of procedures that do this.

We must note that we describe here only a subset of the possible approaches available. Henry and Gates (2017) provide an overview of search approaches that have been used in functional MRI, but can be applied to any time series data collected in studies focused on human psychology, behavior, or physiology. However, many approaches require a far greater number of time points than is typically available. In these cases, most developers recommend concatenating the individuals' data to arrive at one long data set and apply the method to that data (e.g., Ramsey, Hanson, & Glymour, 2011). This is similar to multi-subject approaches that have invariant parameter estimates across subjects. The problem with such approaches is that individual-level nuances are not accounted for by aggregating across individuals (described in detail in Section 6.6). Hence, we focus on methods that explicitly allow for individual-level nuances in the pattern of relations. Throughout this chapter, we utilize the Fisher and Boswell (2016) data first introduced in Chapter 1. We now turn to the three main classes of model-building approaches: filter, wrapper, and embedded methods.

6.2 Filter Methods

Filter methods utilize statistical tests to identify which features to keep outside of any indication of performance. They are a common technique for model modification in the social sciences even if they are not typically called by this term. For example, in functional MRI, researchers conduct a large number of *t*-tests on brain region data to identify if there were a significant increase or decrease in the brain activity indices between tasks. From here, a portion of the t-tests is discussed in papers (i.e., the ones that met statistical significance after controlling for multiple comparisons). Another example would be if a researcher arrived at correlation coefficients for a large number of variables and rank-ordered the values in order to select a prespecified number of values according to the absolute values of the coefficient estimates. In this case, the variables related to the high correlation values may be selected for further analysis. Filter methods come in many forms that are dictated by the qualities of the data. What distinguishes filter methods from the methods to follow is that the objective criterion is not related to performance in a machine learning sense. Namely, multiple models are not tested and compared to arrive at the set of features.

6.3 Wrapper Methods

The most common model selection approaches used in the social sciences fall under the wrapper methods category. Wrapper methods compare competing models that have subsets of possible relations and sequentially add or remove relations in the model. Many readers may have already utilized wrapper approaches. Perhaps the most commonly used is forward selection hierarchical regression, where potential variables are added to see if the newly added variable (or set of variables) significantly improves the amount of variance explained by the model. Its complement, backward-deletion hierarchical regression, is another type of wrapper method that may be familiar to readers. Wrapper methods were used in Chapter 3 to arrive at relations to include in final factor models, and in Chapter 5 when adding significant paths to our dynamic factor models.

We describe here the foundations of the tests that can be used to add or delete relations based on significance tests: Wald's test, likelihood ratio tests, and score functions. For each of these methods, we rely on the notion of alternative and null models, whereby alternative models represent models where parameters are not constrained to zero, and null models are those in which these are constrained to zero. The Wald approach starts with the alternative models and considers whether constraining a relation to zero offers an improvement in the amount of variance explained by the overall model. Likelihood ratio tests compare both the null and alternative models directly. Finally, the use of score functions—specifically Lagrange multipliers and modification indices (MIs)—start at the null and determines whether removing a restriction (i.e., letting a parameter be estimated in the model) provides significant improvement. Much of the details regarding these approaches come from Engle (1984). After explaining these broad approaches, we turn to one approach that uses a combination of these for arriving at final uSEM/SVAR models.

6.3.1 Wald's Test

Wald (1943) formalized the notion of comparing an estimated parameter $\hat{\theta}$ to some null value for that parameter (typically zero, denoted here as θ^0) to identify if the relation should be included in the model. Formally, the difference between the estimated and null $\hat{\theta} - \theta^0$ can be evaluated using the familiar t-statistic used to evaluate parameter significance in the regression context:

$$t = \frac{\hat{\theta} - \theta^0}{SE_{\hat{\theta}}} \tag{6.1}$$

where $\theta^0 = 0$ and $SE_{\hat{\theta}}$ is the standard error of the estimated parameter. The t-statistic can be evaluated for significance with absolute values over 1.96 being considered significant in a two-tailed test given an alpha of 0.05, assuming the sample size is relatively large.

Wald tests can be utilized for completely data-driven searches by starting with a model where all candidate parameters are estimated, and iteratively removing each relation that has (1) the lowest *t*-value and (2) is not significant after controlling for the other variables currently in the model. Another option is to conduct a semi-confirmatory search where the model includes parameters that are of specific interest to the research based on hypotheses. From this starting point, modification to this model is made by deleting parameters from the model that are not significant and not of interest to the researcher. In both exploratory and semi-confirmatory cases, the approach is analogous to backward deletion in a regression sense. In stepwise searches such as this, the final model will typically be sensitive to which variables are removed as predictors in the early steps. This approach is not possible in cases where the saturated model cannot be identified, such as in uSEM.

6.3.2 Likelihood Ratio Tests

Likelihood ratio tests compare the likelihood in one model with the likelihood obtained in a nested model. A *nested model* has the same relations as the competing model except it has some sort of restriction, or set of restrictions, that do not exist in the unrestricted (full) model. These restrictions can be either (1) the omission of one of more relations (i.e., set a parameter to equal zero) or (2) setting two parameters to be equal that are allowed to be estimated in the full model. In factor analysis, comparing a two-factor model that has all possible cross-loadings (except the scaling indicators) with a model that contains only a portion of these would represent a full and nested model, respectively. When only one relation is tested at a time the result will likely be very similar to the Wald test introduced above.

Chapter 3 describes how, in SEM, the likelihood of obtaining the model-implied covariance matrix compared to the overall model fit statistic for a saturated model is a likelihood ratio, which is asymptotically chi-square distributed. In that case, we test if the likelihood of the nested model (which uses a subset of possible parameters) is significantly different than the covariance matrix obtained from the observed data. The test thus compares the model-implied covariance matrix with the observed covariance matrix, which serves as an ideal null model in this case as it represents a perfect fit (Bollen, 1989, p. 265). We can similarly arrive at differences in fit between any two models where one is nested within another.

Following Bollen (1989, p. 292), the likelihood ratio test, or χ^2 difference test, among nested models in SEM can be defined as

$$\chi^2_{diff} = -2\left[logL\left(\widehat{\Theta_r}\right) - logL\left(\widehat{\Theta_u}\right)\right]$$

where $logL\left(\widehat{\Theta_r}\right)$ is the log likelihood of the restricted (nested) model and $logL\left(\widehat{\Theta_u}\right)$ the log likelihood of the unrestricted model. The χ^2_{diff} test statistic is χ^2 distributed[1] with degrees of freedom equal to the difference in the degrees of freedom for the two competing models. One can see if two models are significantly different in their fit, and if so, select the better-fitting model.

6.3.3 Score Functions

Score functions provide a flexible and efficient method for model searches because they do not always require that parameters of the alternative (e.g., less restricted) model be estimated. The Lagrange multiplier (LM) and its equivalent, the modification index (MI; Sörbom, 1989), allow for the statistical test of whether the addition of a parameter to a given model will improve the likelihood of the model while considering parameters that are already estimated in the model. Maximizing the log-likelihood to the constraint $\Theta_u = \Theta_r$, where Θ_r is typically set to zero, enables the identification of the cost of this constraint. High costs suggest that the constraint may be inconsistent with the data. We start by first defining the score as

$$s(\Theta_r) = \partial logL(\Theta_r) / \partial\Theta_r$$

Which provides the slope of the log likelihood, given a change in the parameter value from the restriction (Θ_r, usually zero) to unrestricted (Θ_u, estimated). The statistic:

$$LM = s'(\Theta_r)' \, V(\Theta_r)^{-1} s(\Theta_r) / T$$

where $V(\Theta_r)^{-1}$ is the variance of the score divided by T. LMs will have a central χ^2 distribution with degrees of freedom equal to the difference in the number of parameters estimated in the two sets θ_u and θ_r. If the LM is high, the constraint that the unrestricted model equals the restricted model

[1] This distribution holds if the constraints are tenable.

(where a given parameter is not estimated) is rejected, implying that the parameter should be estimated rather than set to zero.

6.3.4 Example: Automated Relation Selection Using Wrapper Methods

6.3.4.1 *Model Search Procedure*

An exemplar approach introduced in 2010 (Gates et al.) utilizes a combination of wrapper methods to arrive at the structures of relations in individual-level unified SEMs (uSEM) in a data-driven fashion. Of note, this approach forms the foundation for the original GIMME method described in Section 6.7. uSEMs are one way to estimate SVAR models (described in Chapter 4) which for a lag of one is

$$\mathbf{y}(t) = \mathbf{A}\mathbf{y}(t) + \mathbf{\Phi}_1 \mathbf{y}(t-1) + \epsilon(t)$$

where $\mathbf{y}(.)$ is a p x 1 vector of values at the time point indicated in the parentheses across all possible t, $\mathbf{A}$ is an asymmetric p x p matrix containing estimates for the contemporaneous relations among the $\mathbf{y}(t)$ variables on the off-diagonal, $\mathbf{\Phi}_1$ an asymmetric pxp matrix containing the AR effects on the diagonal and cross-regressive coefficient estimates on the off-diagonal, and $\epsilon(t)$ the innovations or equation errors at each time point. For uSEMs, equation errors are assumed to be uncorrelated with each other and across time.

To estimate these with SEM, we first time-embed the variables such that the first set of variables are the p variables at $t-1$ and the next set the same variables at t. In Chapter 5, we presented the **B** matrix used in dynamic factor models with lagged relations among latent variables. Here, we slightly alter the **B** matrix structure to allow for contemporaneous relations:

$$\boldsymbol{B} = \begin{bmatrix} 0 & \cdots & \cdots & \cdots & 0 & 0 & \cdots & \cdots & \cdots & 0 \\ \vdots & \ddots & & & \vdots & \vdots & \ddots & & & \vdots \\ \vdots & & \ddots & & \vdots & \vdots & & \ddots & & \vdots \\ \vdots & & & \ddots & \vdots & \vdots & & & \ddots & \vdots \\ 0 & \cdots & \cdots & \cdots & 0 & 0 & \cdots & \cdots & \cdots & 0 \\ \phi_{11} & \cdots & \cdots & \cdots & \phi_{1p} & 0 & a_{12} & \cdots & \cdots & a_{1p} \\ \vdots & \ddots & & & \vdots & a_{21} & \ddots & \ddots & & \vdots \\ \vdots & & \ddots & & \vdots & \vdots & \ddots & \ddots & \ddots & \vdots \\ \vdots & & & \ddots & \vdots & \vdots & & \ddots & \ddots & a_{(p-1)p} \\ \phi_{p1} & \cdots & \cdots & \cdots & \phi_{pp} & a_{p1} & \cdots & \cdots & \cdots & 0 \end{bmatrix}_{2p \times 2p}$$

Here, the **A** and $\mathbf{\Phi}_1$ matrices are submatrices of this larger **B** matrix. The non-zero elements of this matrix indicate all candidate paths, or relations, among variables that can be included for a uSEM. However, in practice, we must constrain some of these to zero for identifiability purposes, and sparsity is also preferred to ease interpretation. We now turn to the automated procedure.

The approach described here operates from within a Granger causality perspective, which exists if, after controlling for autoregressive effects, variability in a given variable can be explained by another variable at a lag or contemporaneously (Granger, 1969). Another important feature of the approach is the use of LM equivalents called MIs (Sörbom, 1989). Within the SEM framework, MIs can be generated for each potential path that is a candidate for addition to the model. In this case, there is only one degree of freedom difference from the sets θ_r and θ_u (restricted and unrestricted, respectively). Each parameter can be evaluated since it is χ^2-distributed to see if it would explain a significant amount of variability in the data if freed for estimation given the current model's parameter estimates.

The algorithm performs as follows (see Figure 6.1 for a depiction). First, a baseline model is estimated with only autoregressive parameters estimated (the restricted model with parameters θ_r). Second, MIs of all candidate lagged and contemporaneous paths are evaluated (i.e., non-zero elements in the **B** matrix above), with the largest one that is statistically significant (after a strict Bonferroni correction) selected and added to this baseline model. Step two iterates until the model obtains an acceptable fit. Specifically, the

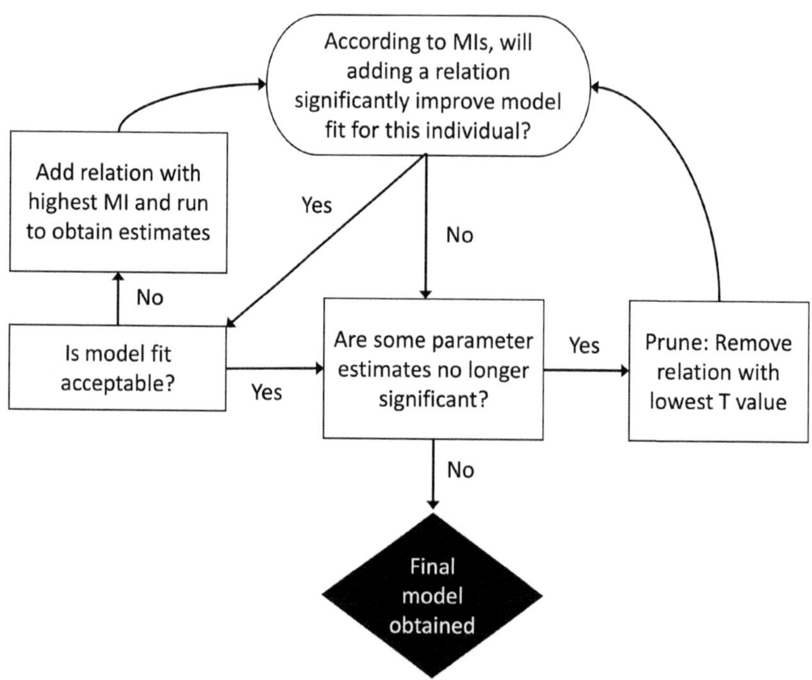

FIGURE 6.1
Automated individual-level uSEM heuristic.

entire model fit is evaluated to assess whether the search should continue or stop. Four fit indices are evaluated after the addition or removal of each path: Tucker Lewis index, Non-Normed Fit Index, Root Mean Square Error of Approximation, and Standardized Root Mean Square Residual (see Brown, 2006). This addition of paths continues until any two out of the four fit indices suggest an acceptable fit. Using fit indices as a stopping mechanism rather than continuing to add potentially significant paths favors parsimony in the model-building approach. Finally, Wald's tests are conducted for each model to evaluate if a path in the model has become non-significant after adding other paths. Autoregressive parameters are not removed even if they are not significant.

6.3.4.2 Simulated Data Example

For this example, data were simulated for $N = 20$ individuals with $T = 200$. The mean parameter specifications were

$$A = \begin{bmatrix} 0 & .5 & 0 & 0 & 0 \\ 0 & 0 & 0 & 0 & 0 \\ 0 & .4 & 0 & 0 & 0 \\ .5 & 0 & 0 & 0 & 0 \\ 0 & 0 & 0 & .6 & 0 \end{bmatrix} \quad \boldsymbol{\Phi} = \begin{bmatrix} .5 & 0 & 0 & 0 & 0 \\ 0 & .5 & 0 & 0 & 0 \\ 0 & 0 & .5 & 0 & 0 \\ 0 & 0 & 0 & .5 & 0 \\ 0 & 0 & 0 & 0 & .5 \end{bmatrix}$$

with parameters for each individual drawn from a normal distribution (standard deviation of 0.2) and errors generated to be normally distributed and uncorrelated with each other with a mean of 0 and standard deviation of 1. Readers are encouraged to follow along with the online supplement for this chapter (link to code available at https://www.routledge.com/9781482230598), where you can generate this data and conduct the analysis to obtain the results we depict below.

We begin by estimating the baseline model for one individual with only the autoregressive parameters estimated. From here, we can explore the MIs that are obtained. To build intuition, we provide Table 6.1, which is the first five rows of candidate paths that could be added from the baseline model. We see that the MIs are rather low until we get to the MI relating to the regression coefficient in the regression of V1 on V2. This parameter, indicated a_{12}, is non-zero in the data generating model. From this small example, we can already see how well the MIs perform in identifying which coefficient, if freed, will likely have a significant estimate. The largest MI in this example data set is V5 regressed on V4, which corresponds to the largest contemporaneous parameter that exists in the **A** matrix as specified in the data generating model.

Model revision using the MIs could be carried out manually by iteratively selecting a path to free for estimation, running the model with that path added, evaluating all paths for significance, pruning, and stopping

TABLE 6.1

Example MI Values Obtained from Baseline Model.

RHS	Op	RHS	MI
V1	~	V2lag	8.73
V1	~	V3lag	7.87
V1	~	V4lag	0.12
V1	~	V5lag	0.40
V2	~	V1	45.75

Note: Higher values are expected to be associated with greater improvement to the fit of the model should this relation be included. "RHS" and "LHS" denote the right- and left-hand side of the regression equation, respectively; "Op" indicates which operator, which in *lavaan* and LISREL syntax is "~" for regression paths.

once the model evidences acceptable fit. A quicker option is to utilize the *indSEM* function within the *gimme* R package, which conducts the algorithm described here automatically. The results from that approach are summarized in Figure 6.2. Looking across all individual-level results (Figure 6.2A) the path structure seems to be recovered fairly well for most individuals as there are no spurious paths. However, from the variability in line thickness, we can see that some paths were missed for a portion of the individuals.

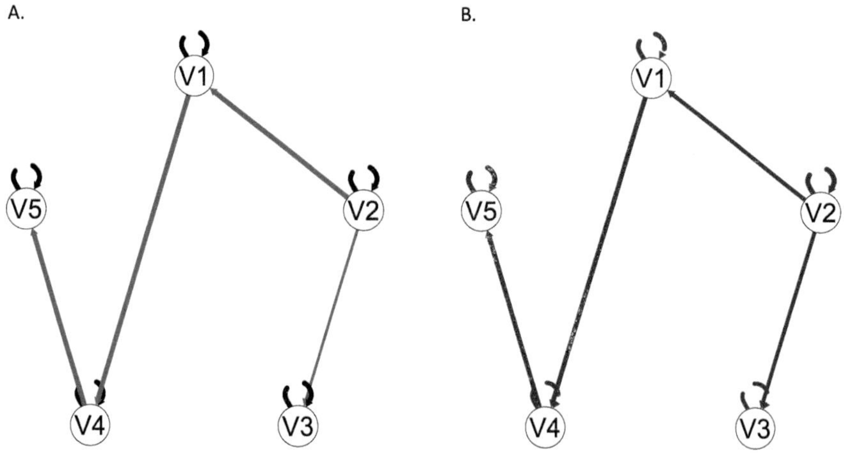

FIGURE 6.2

Results from data simulated for multiple subjects with identical data generating models. All model searches were conducted at the individual level using the *indSEM* function. Panel (A) aggregates the results across individuals, with path width indicating the proportion of individuals with that path. Black paths occurred for 100% of the individuals; grey paths were found only for a subset of individuals. Panel (B) depicts results from one of the data sets. Here, large line widths indicate higher absolute values of estimates. Red (hot) reflects positive values, blue (cold) reflects negative values.

6.3.4.3 Empirical Data Example

The automated uSEM approach described above has been used in neuroimaging (e.g., Hillary et al., 2011), behavioral observations (e.g., Beltz et al., 2013), and ecological momentary assessment (e.g., Fisher & Boswell, 2016) data. Here, we demonstrate its use on one individual from the data made publicly available by Fisher and Boswell (2016). We see here results from the combination of wrapper methods just described. Using the *indSEM* function in the *gimme* R package (Lane et al., 2022), this method found most relations among variables to be contemporaneous (see Figure 6.3), suggesting that the temporal resolution of the data acquisition (every few hours) was larger than the process being studied. For example, Positive and Energetic are positively and contemporaneously related, suggesting that this person feels the emotions in tandem, or that any lead-lag relations among Positive and Energetic are occurring faster than the rate of the individual responding. Interpretation is similar to that in cross-sectional studies: after taking into account covariates, having information on Positive assists in explaining the variance in Energetic. The reverse is not seen here. Different designs and data qualities may reveal different types of relations, such as more lagged relations.

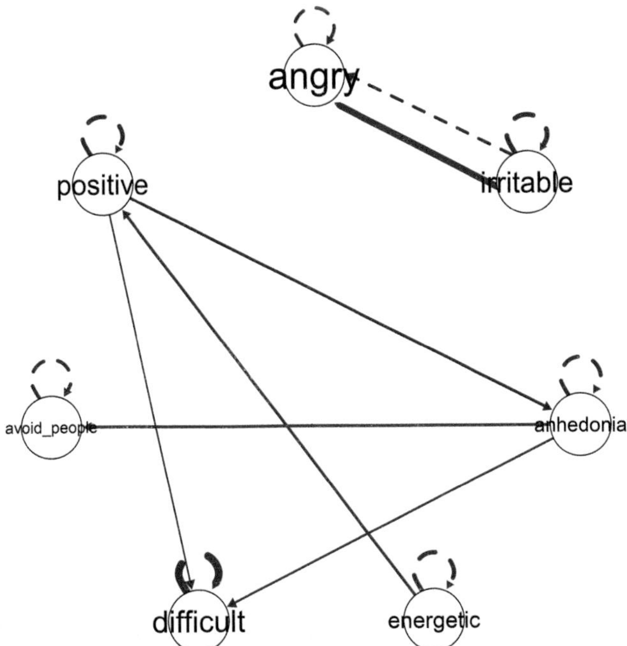

FIGURE 6.3
Exemplar individual from the Fisher and Boswell (2016) ecological momentary assessment data. Following heatmap convention, red (hot) lines represent positive values and blue (cold) lines represent negative relations. Solid lines represent contemporaneous relations and dashed lines lag-1 relations.

6.3.5 Conclusion on Wrapper Approaches

Wrapper approaches carry a number of benefits. For one, most researchers are already familiar with them, thus aiding in the appropriate execution of the analyses, and interpretation of the results. Two, they are quite flexible. The researcher can easily choose to override a given test, and include a given parameter relating two variables even if it did not pass the significance test. These approaches can be used for minor model modifications in a confirmatory setting or used in a completely exploratory, data-driven manner. Three, with enough power, which usually is ascertained with simulation studies, these methods usually do a good job of recovering the true pattern of relations.

One note of caution is that a percentage of significant results may be due to chance. This issue can be assuaged by using stricter values when testing for significance. Familywise error rates, Bonferroni corrections, and the use of false discovery rates all can be explored as possible solutions if the number of tests is large.[2] Another commonly employed technique is to favor parsimony in the final model. This is utilized in the automated uSEM approach by stopping the model search if the fit indices suggest adequate model fit. That is, even if additional relations might be significant, they are not added in favor of parsimony.

A primary drawback of iterative approaches is the step-wise nature. In some cases, the final solution can be greatly influenced by which relation was chosen first (Beltz & Molenaar, 2016; Weigard et al., 2021). In fact, a large-scale simulation study revealed that while *indSEM* can recover the presence of a relation between two variables, correct identification of the directionality of the relation can be poor (Gates & Molenaar, 2012). This is likely due to the stepwise nature of the approach and possibility of early paths being selected due to noise in the individual data. This drawback has given rise to the increased popularity of approaches that avoid step-wise procedures, which we now turn to.

6.4 Embedded Methods: Regularization

Embedded methods present a natural complement to the prior two approaches. Filter methods select relations with no learning, whereas wrapper machine learning methods select the best subset of features by measuring their relative contributions to explaining variance in the data. In embedded methods, the learning occurs simultaneously with estimation. Commonly used examples in psychology are decision trees (Quinlan, 1986), Least Absolute Shrinkage Selection Operator (LASSO; Tibshirani, 1996), and elastic net (Zou & Hastie, 2005). Decision trees can be used to predict categorical or continuous values

[2] In the individual-level search procedure introduced here, Bonferroni corrections were used.

of a single outcome variable. Since most analyses and research questions typically consider multiple dependent variables in time series analysis, the present description focuses on LASSO and elastic net, which can accommodate the learning of model structures of models discussed in this book such as VAR, P-technique, dynamic factor analysis, and uSEM.

To begin, we take the simplest case for pedagogical purposes—that of a linear regression. Assuming an intercept of zero, those familiar with linear regression will recognize that the cost function that is minimized is

$$\sum_{i=1}^{N}\left(y_i-\sum_{j}\beta_j x_{ij}\right)^2.$$

The terms within the parentheses are the regression residuals, obtained after subtracting from the observed y_i value the predicted value obtained by summing the product of the $j = 1, 2, \ldots p$ β times the respective predictor x variable for individual i. The goal is to minimize the squared errors.

Simultaneous estimation and variable selection is conducted by minimizing:

$$L(\lambda_1,\beta)=\sum_{i=1}^{N}\left(y_i-\sum_{j}\beta_j x_{ij}\right)^2+\lambda_1\sum_{j}\left|\beta_j\right|.$$

The λ_1 parameter is the regularization penalty. As λ_1 increases, the regression coefficients are penalized more, resulting in smaller ("shrunken") estimated values for the β coefficients. When $\lambda_1=0$ the estimates are the same as least square estimates as there is no penalty. Those familiar with ridge regression may see some similarities in the equations. The difference is that LASSO uses an $\ell 1$-norm penalty whereas ridge regression uses the $\ell 2$-norm penalty by squaring the terms within the summation on the right rather than taking the absolute values. LASSO has gained popularity over ridge regression by shrinking some parameters to zero, thus making it an ideal approach for feature selection in a data-driven manner.

Elastic net includes both ridge regression and LASSO penalties by minimizing:

$$L(\lambda_1,\lambda_2,\beta)=\sum_{i=1}^{N}\left(y_i-\sum_{j}\beta_j x_{ij}\right)^2+\lambda_1\sum_{j}\left|\beta_j\right|+\lambda_2\sum_{j}\beta_j^{\,2}.$$

A benefit of the elastic net (and ridge regression) over LASSO is that, if multiple variables are correlated, the coefficients of correlated predictors are shrunk toward each other. If the goal is to retain all variables that relate to a given outcome, then this may be preferred over LASSO, which tends to just choose one of the variables in selecting predictors. For this reason, elastic net has become the preferred method for simultaneous variable selection and estimation. Simulation studies suggest that elastic net outperforms in terms of recovering the true effects that relate to the outcome (Ogutu, Schulz-Streeck, & Piepho, 2012; Waldmann et al., 2013; Zou & Hastie, 2005).

A popular package for conducting regression, LASSO, and elastic net is *glmnet* (Friedman, Hastie, & Tibshirani, 2010). Here, the user can define which approach they would like to use for regularization by setting what they call an elastic net mixing parameter, α, *a priori*. The impact of alpha can be seen best in equation form:

$$L(\lambda_1,\lambda_2,\beta)=\sum_{i=1}^{N}\left(y_i-\sum_j \beta_j x_{ij}\right)^2+\lambda\left[(1-\alpha)/2\sum_j \beta_j^2+\alpha\sum_j\left|\beta_j\right|\right].$$

Note the presence of only one penalization parameter (λ). Setting α to zero returns us to a ridge regression by retaining only the portion in brackets that relates to the $\ell 2$-norm of the estimates. An α of 1 does the opposite, removing the $\ell 2$-norm portion and retaining only the $\ell 1$-norm, thus returning us to LASSO. According to the developers, setting α to 0.5 provides an elastic net penalization.

Choosing the value for the penalization parameter λ represents a key aspect of regularization. As with other approaches introduced in this book, fit indices can be used to assess the best model fit. Fit indices such as the BIC or AIC can be used by obtaining these values for models fit across the span of λ and identifying the model with the lowest fit value. Another commonly used approach is *k*-fold cross-validation. Here, the data are randomly placed into k non-overlapping folds (groups) of equal size. Fold sizes of $k = 10$ or 20 are commonly used. To start, one fold is left out and regularization is conducted on the combined remaining $k - 1$ data (the "training" set). The model obtained is then fit to the left-out fold and the mean squared error (MSE) is obtained. This process iterates throughout each of the folds to obtain an average MSE at that λ to identify at which λ the MSE is the smallest. We demonstrate the use of regularization with a popular R package, *graphicalVAR* (Epskamp, 2017), commonly used in the social sciences.

6.4.1 Exemplar Approach: Regularization in Graphical VAR

The *graphicalVAR* package (Epskamp, 2017), obtains sparsity in the VAR model using LASSO with the BIC or extended BIC for model selection (Epskamp et al., 2018). To induce sparsity in the variance/covariance matrix of residuals from this VAR model, the standardized precision matrix (the inverse of the covariance matrix) of the residuals is also subjected to LASSO penalization. Standardizing the precision matrix provides the partial correlation of each residual and is referred to as the Gaussian Graphical model when the data have a multivariate normal distribution. The contemporaneous relations are the partial correlations among residuals following VAR analysis and controlling for all other residuals' potential influence on each given pair of residuals. These paths are often denoted as "undirected" since they are on the inverted covariance matrix. VAR models can be depicted graphically with nodes (vertices, variables) and edges (arrows, paths) much like SEMs (Dahlhaus & Eichler, 2003; Lauritzen, 1996). This graphical representation gave rise to the graphical VAR term.

As an example of a regularized approach, we conducted graphical VAR on the same data used in the feed-forward uSEM example (Figure 6.4). Overall, we see a lot of consistencies. Similar to the uSEM results, there was a strong positive contemporaneous relation between Angry and Irritable. There also are similar paths among Positive with Anhedonia (negative sign) and Energetic (positive). Anhedonia was contemporaneously related to Social Avoidance in both sets of results. One difference between the two is that the graphical VAR results found the relation from anhedonia too difficult to be lagged, whereas in uSEM this was found to be contemporaneous. As seen in Chapter 4, data-generating contemporaneous directed relations may surface as lagged relations among observed variables if a standard VAR is conducted (as is done here). As a uSEM can be transformed to an equivalent VAR, this does not mean the relation is spurious. Rather, it is representing relations in a different way, and as such interpretation of these effects varies greatly. In the literature, lagged relations are often interpreted as causal since

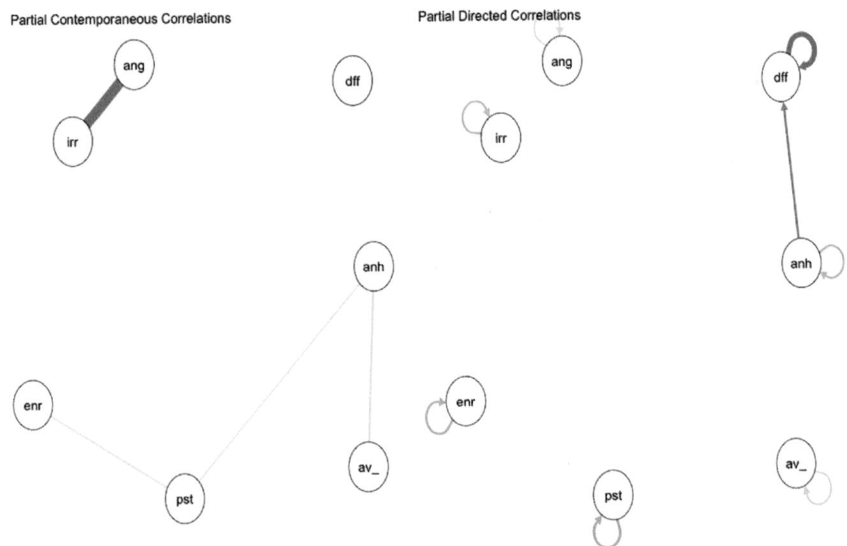

FIGURE 6.4
graphicalVAR results for Fisher example. Partial directed correlations represent the lag-1 relations among variables. The partial contemporaneous correlations graph depicts the partial correlations among the residuals following VAR analysis. Red indicates negative, green indicates positive. "Ang" = Angry; "irr" = Irritable; "anh" = Anhedonia; "enr" = Energetic; "diff" = Difficult; "av_" = Avoid people; "pst" = Positive.

temporal precedence exists. Given that this is an empirical example the data generating mechanism is not known as it is in simulated data. The similarity in results provides convergent evidence regarding the relations among the variables for this particular person.

6.5 Problems with Individual-Level Searches

Conducting model searches from a completely null model historically has been ill-advised. The possibility of noise driving the model search is always a potential issue when attempting to recover individual-level IAV patterns of temporal relations. In a seminal paper evaluating 26 different methods for recovering directed relations among time series variables across 28 different simulation conditions, Smith et al. (2011) reported that none of the approaches can both recover the (1) presence of a relation and (2) direction of the relation at rates higher than chance. Included in the methods tested were a number of variations of Granger causality from within the familiar VAR framework and directed acyclical graphs. SEM approaches were not tested due to a lack of a publicly available data-driven approach. Gates and Molenaar (2012) investigated this data with the automated search for uSEM patterns described above and found that while the presence of a relation between two variables could be detected with this approach, the directionality of the relation could not be correctly identified at rates greater than chance.

This simulation study highlighted the potential for person-specific models to provide patterns of relations that do not correspond with the data generating process. One potential reason could be that the models are overfit, meaning that the model has more parameters than needed to explain variability in the process of interest. This often occurs because the residual or noise variance is also being explained by the model. A consequence of overfitting is that the model does not fit new data well. In the context of person-specific time series analysis, this implies that there may be spurious relations which will have coefficients that tend toward zero once updated with new or incoming data for that person.

Utilizing data across individuals when discovering the model structure may help to buffer against overfitting by picking signal from noise by looking across individuals in a sample. Two primary methods exist for arriving at sets of relations that may replicate across individuals: (1) aggregating data sets to arrive at one composite data and conducting analysis on this; and (2) selecting relations that are consistently found for the majority of the sample and estimating those for all individuals. The former approach has been implemented in a method for pooled graphical VAR (function *mlGraphicalVAR* in the *graphicalVAR* package) and the aggregated uSEM approach available in the *gimme* package (function *aggSEM* in the *gimme* package).

The selection of relations that replicate across individuals has been used for uSEM in Group Iterative Multiple Model Estimation (GIMME; Gates & Molenaar, 2012), which is the primary and recommended function available in the *gimme* R package.

After finding the pattern of relations that is thought to exist across individuals (i.e., group-level relations), these methods can then use that as a foundation model structure from which to search for the individual-level relations. This approach is done within the GIMME algorithm. By using information gained by examining data across individuals, the individual-level search is starting closer to the final model. We next provide details and examples of the aggregation and replication approaches.

6.6 Data Aggregation Approaches

Aggregating data by appending individuals' time series data into one matrix is a very common approach for attempting to arrive at a model that describes the "average" individual. Here, the resulting matrix is of dimension $\Sigma_{i=1}^{N} T_i$ *by* p, where T_i is the number of time points for individual i and p is the number of variables. In functional MRI studies, it is common to concatenate individual data to arrive at one correlation matrix that is thought to describe typical functional connectivity of the human brain. As one example, this approach is done for IMaGES (Ramsey et al., 2011), with developers of this algorithm indicating that combing data sets is often required to obtain the necessary number of time points. Researchers using data collected across days may be tempted to aggregate data in this manner to increase power and the number of observations in the analysis. It must be stressed that the assumption when conducting analysis on aggregated data is that all individuals are identical in their processes. This assumption of ergodicity, as defined in Chapter 2, is likely not realistic. Given the popularity of these methods, we first describe them here to provide context before discussing some issues that arise should the assumption of homogeneity in individuals be violated.

6.6.1 Exemplar Output of Aggregated Approaches

Pooled graphical VAR (function *mlGraphicalVAR*) provided within the R package *graphicalVAR* and aggregated uSEM (function *aggSEM*) in the *gimme* package arrive at group-level or fixed results by pooling the data across individuals. This is done by concatenating data such that one person's time series is appended below another until all individuals are in one large data frame. Importantly, individuals are mean-centered. In this way, deviations from the grand mean will not influence results. The *graphicalVAR* function described above is then applied to this data in *mlGraphicalVAR* (in the *graphicalVAR* package), and the *indSEM* function described previously is applied to this data set for *aggSEM* (in the *gimme* package). Additionally, individual-level analyses can always be carried out using the *graphicalVAR* or *indSEM* functions on each individual. Of note, the individual-level parameters and pattern of relations is not constrained in any way to those obtained in the fixed effect approach. This differentiates these approaches from traditional multilevel modeling, which takes into account group-level estimates.

Figure 6.5 depicts the VAR and uSEM results from the aggregated data when using the entire sample of individuals in the Fisher data set. The results are largely similar across the two methods. Very few lagged relations were retained in the gVAR results, and those that were had very low coefficient estimates (see *Estimated fixed PDC*). This is similar to the uSEM results, where no cross-lagged relations were found. For the contemporaneous effects, which are the partial correlations of the residuals following the VAR analyses in gVAR, two edges with high coefficient values, as evidenced with thicker paths, were recovered: the relations between *Anger* and *Irritability*, and between Positive and *Energetic*. These relations also existed in the uSEM results in Panel B. One major difference between the two approaches is that the gVAR method returned a greater number of relations than the uSEM method.

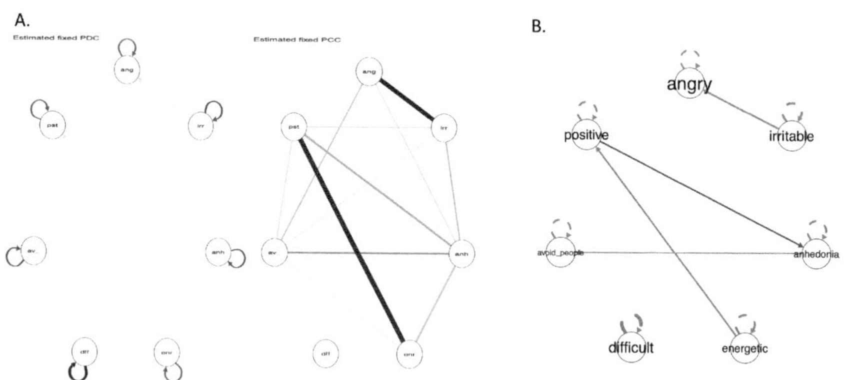

FIGURE 6.5
Aggregated data results. Results from gVAR (Panel A) and automated uSEM (Panel B) conducted on data that were aggregated across individuals. In Panel A, red indicates negative and green positive. In Panel B, red indicates positive and blue indicates negative following heatmap convention. These figures were obtained by using *mlgraphicalVAR* and *aggSEM* functions in *graphicalVAR* and *gimme* packages, respectively.

6.6.2 Issues with Traditional Forms of Aggregation

Before moving on to replication approaches, it is important to explain the drawbacks of aggregating, or pooling, data to arrive at the group-level model.

Researchers may believe aggregating data arrives at an average model the covariance matrix obtained from aggregated data set would approximately equal the average of the individuals' covariance matrices. However, this approach has some problems. The main one is that the approach assumes ergodicity exists across individuals. This is not always the case. A second problem is that the results may be swayed by outliers. For example, if a few people in the sample have a very strong relation between two variables, that would make that relation more likely to be selected even if others in the sample do not have this relation. A third problem is that relations that truly do exist for everyone in the sample might not be retained in the model due to sign differences. In a boundary case where all individuals have a strong relation between a given pair of variables, the relation might not be selected if it is negative for half the individuals and positive for the other half. A final problem is that in the presence of heterogeneity spurious results may appear due to similar partial correlations resulting from the differing patterns.

We present a simulated data example in Figure 6.6 (code provided online) that demonstrates the issue of aggregation. Here, two individuals were simulated to have nearly identical patterns of relations. Only one aspect of the pattern differed: for one individual, Variable 4 explained variance in Variable 1, whereas for the second individual, the direction of the relationship was reversed. The two individuals had identical data generating parameters, except for the path between Variables 1 and 3. For demonstration purposes here, we utilized the *aggSEM* function in the *gimme* package, which concatenates the data such that the time series data for Individual 1 was appended above Individual 2's time series data and conducts the data-driven search described for automated uSEM (Section 6.3.4). Note that the *aggSEM* procedure is not recommended, and is used here solely for pedagogical purposes.

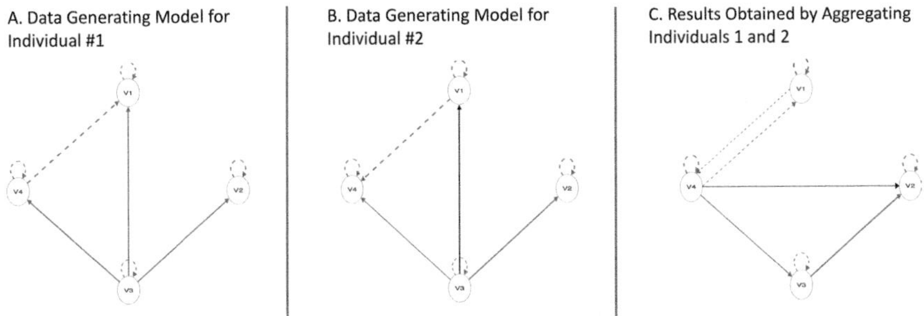

FIGURE 6.6
Example of problems with aggregating data. Data for two individuals were generated to be very similar except the directionality of one path and the sign of another (Panels A and B). Panel C depicts results obtained when the data are concatenated and treated as one data set using the *aggSEM* function in *gimme*. Red lines are positive; blue negative; line width corresponds with parameter weight.

A number of issues arise. First, a spurious relation surfaced between Variables 2 and 4. Second, the relation between Variables 1 and 3 was not found even though it existed for both individuals (albeit with different signs). Finally, both of the directions for Variables 1 and 4 were recovered. However, interpretation of this might suggest that there is a recursive or bidirectional relation between these two variables. In reality, we know that this is not true, given the data generating models. This toy example points to the potential problems that even a small amount of heterogeneity can cause when data are aggregated.

Conducting analyses on data aggregated across individuals requires a strong assumption that the individuals are identical in patterns and weights. Any differences can potentially cause spurious findings and misguided inferences. The issue will be exacerbated with more individuals as they will likely introduce more idiosyncrasies. For these reasons, aggregating individuals as though they are the same is not advised. We now turn to GIMME, which does not aggregate individuals prior to analysis but rather analyzes each individual separately.

6.7 Replication Approaches: Group Iterative Multiple Model Estimation (GIMME)

The issues in correctly recovering the patterns of relations in individual-level models lead to the development of GIMME (Gates & Molenaar, 2012). In GIMME approaches, each individual is considered a unique sample much like in individual-level analysis. To avoid the problems seen in aggregated data, the GIMME algorithm looks at individual data sets across the sample (i.e., group) to identify consistencies across individuals to detect signal from noise. This shared information is then used to prime the search for individual-level relations, thus following suggestions of getting the model closer to the final model before searching (MacCallum, Roznowski, & Necowitz, 1992). The identification of which relations are consistent across the group is analogous to meta-analysis, where results obtained from different samples are examined to identify which results replicate. Figure 6.7 depicts the GIMME heuristic.

The concept is rather straightforward and flexible enough to be applied to various estimation and search approaches as well as various models (e.g., VAR, SVAR/uSEM, or a hybrid of these two). The key component that defines GIMME is the search for relations that *replicate* across individuals. This is done by detecting which relations exist for the majority of individuals in the group by analyzing each individual's data separately and independently of the others. Importantly, the detection of a pattern of relations that exists for the majority of individuals does not do so by aggregating the data in any way.

Step 1: Identify Group-Level Relations

Identify relations (or network edges) that exist for the majority of individuals by looking for replicability across individuals

Step 2: Identify Individual-Level Relations

Use the group-level pattern of relations as a starting point for the search for individual-level relations

FIGURE 6.7
GIMME heuristic. These two steps comprise the underlying method for learning group and individual-level patterns of relations.

6.7.1 Original GIMME

The first use of the GIMME algorithm operated from within a uSEM framework using MIs to guide the search. We focus on the original implementation here as a foundation. Similar to the automated individual-level search procedure introduced in Section 6.3.4, the original GIMME utilizes MIs for selecting relations, Ward's test for pruning relations, and fit indices derived from likelihood ratio tests as a stopping mechanism. The difference here is that the search utilizes shared (group) information in the iterative search.

By default, the GIMME search algorithm begins with only the autoregressive effects estimated for all individuals. This follows from the notion of Granger causality introduced in Chapter 4. With the AR coefficients estimated, the algorithm inspects the MIs for all candidate paths across all individuals and the number of individuals for whom each candidate path is significant. A strict Bonferroni correction for multiple comparisons is used where the α is set to $\alpha = 0.05/(q * N)$, where q is the number of unique elements in the sample covariance matrix and N is the number of individuals.

The original GIMME identifies which relation has the highest count of significant MIs when looking across individuals.[3] This relation can be in either

[3] If there is a tie for the highest MI count, then the relation with the highest average MI is selected from the tied candidate paths.

the **A** or the **Φ** matrix. If the count is higher than expected by chance given the power to detect effects, then the path is added for all individuals. These paths are referred to as "group-level paths" since they are estimated for all individuals. The power to detect effects in this context can be difficult to ascertain in new research endeavors. In functional neuroimaging, an extensive simulation study concluded that given the amount of noise and length of the time series seen in typical studies, one can expect to detect around 75% of relations (Smith et al., 2011). Following this insight, GIMME's default for the "group cutoff", or the proportion of individuals who must have the path be significant in order for it to be added at the group level, is set to 75%. Emerging evidence suggests that setting the group cutoff threshold to a more lenient value, such as 51%, might improve upon the recovery of relations without introducing false positives (Gates, Fisher, & Bollen, 2020).

Once the first group-level path is selected, the search continues iteratively until there are no more relations that fulfill the criteria. Since some paths may become non-significant with the addition of new paths, pruning is conducted on the group-level paths. If a specific relation no longer meets the criterion of being significant at the group cutoff threshold (e.g., 75%) level (following Bonferroni correction), then that path is removed and the search for group-level paths conducted once more.

Following the attainment of a group-level structure of relations, the individual-level searches are conducted. The group-level pattern of paths provides the foundational (or null) model for each individual. Importantly, all of the group-level paths are estimated uniquely for each individual. Thus, they can be quantitatively or numerically different even if they are qualitatively the same. The search for individual-level paths commences in a similar way as the automated individual-level search described in 6.3.4. In the end, the algorithm obtains a pattern of relations that were found on the group level and might generalize to the population from which the sample is obtained as well as individual-level patterns of paths, which allow for individual nuances and personalized assessment of each individuals' processes.

Figure 6.8 depicts exemplar results obtained on the full Fisher sample ($N = 40$) using GIMME. The contemporaneous relation between Angry and Irritable, which was quite strong for the individual of the previously used example, was the only path obtained at the group level. Paths among Positive with Anhedonia and Energetic appear to be thicker than other paths, which is consistent with the findings from the aggregated data. Importantly, that these paths were not selected at the group level indicates that they were not significant for the majority of individuals as defined here (75%). The individual-level results, which are the grey paths, suggest that cross-lagged paths exist for some individuals. In the end, it is evident that there is substantial heterogeneity in the patterns of relations among variables.

A number of basic options exist for the user. Users can conduct semi-confirmatory searches by indicating paths that they would like to be estimated for all individuals. A complementary option allows users to indicate paths they would *not* like to be included in the search. Users can also choose to

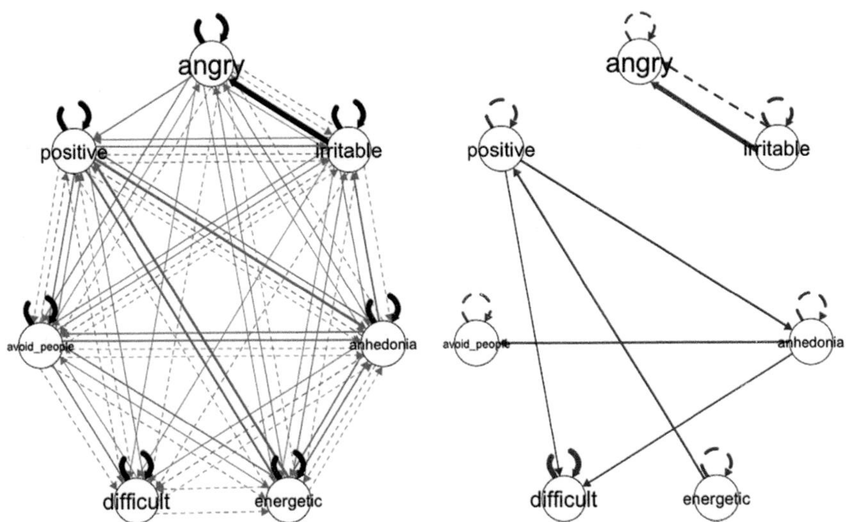

FIGURE 6.8
GIMME results for Fisher example. Solid lines represent contemporaneous (lag-0) relations and dashed lines represent lag-1 relations. Line width corresponds to proportion of individuals with that path. Black lines are group-level paths and were estimated for everyone. For the individual-level result (participant 4, right panel), red lines indicate positive and blue negative, with line width relating to the absolute value of estimate size.

standardize the variables prior to analysis. This is recommended when the variances vary widely across the variables, which may cause issues with convergence. Finally, users can select whether to begin the model search with AR paths freed for estimation at the start of the group-level search. Starting without the AR coefficients estimates takes the search outside of the Granger causality context. In doing so, it is possible that multiple paths may explain equal amounts of variance early in the search procedure.

In addition to these options, a number of extensions exist that change the entire search and estimation procedure. For one, users are recommended to use the *gimme* option that allows for multiple solutions (MS-GIMME) if they choose to not have AR coefficients estimated at the start of the search (Beltz & Molenaar, 2016). Not including the AR coefficients means that the **A** matrix of contemporaneous directed estimates will be symmetrical, making it difficult to ascertain directionality. MS-GIMME would follow the stepwise searches completed for selected one direction as well as, separately, the opposite direction. This results in multiple solutions for the user to choose from (often using AIC).

Two, users can conduct a VAR model search within *gimme* instead of uSEM. Here, no directed contemporaneous relations are allowed in the search space, only the bidirectional covariances among residuals. Three, users can subset individuals into subgroups based entirely on their patterns and estimates (Gates et al., 2017; Lane et al., 2019). This option is explained in detail in Chapter 9 when we discuss community detection of networks. Four, users can search for structural paths in a PFA(1,0) model. Termed latent variable

GIMME (LV-GIMME; Gates, Fisher, & Bollen, 2020), the user specifies confirmatory measurement models. The search finds lagged and contemporaneous relations among the latent variables using the algorithm described above.

Five, if certain variables are exogenous and cannot be logically predicted by other variables in the system researchers can indicate this in the *gimme* arguments. Examples of exogenous variables might be weather, as a person's mood does not predict sunshine, but the reverse can be true, or something manipulated by the researcher such as a task in fMRI studies or treatment. Arizmendi and colleagues (2021) demonstrate how the use of exogenous variables can help to identify treatment effects across time. Duffy and colleagues (2021) introduce a related innovation that convolves a vector indicating the start and end times of an endogenous variable (here, an fMRI task) with a person-specific response function to see the direct, indirect, and modulating impact of this exogenous variable. Finally, there is the option for including in the search space correlations among residuals (as in VAR and graphical VAR) in addition to contemporaneous relations among observed variables (as in uSEM/SVAR). This provides a hybrid-uSEM model that does not require the user to make an assumption regarding which type of contemporaneous relation best matches the data (Molenaar, 2017). We now turn to this important innovation before concluding this chapter.

6.7.2 Hybrid GIMME

The GIMME approaches described thus far only search for directed contemporaneous relations among variables in the uSEM case or correlated errors in the VAR usage. The interpretation of the contemporaneous directed relations is rather straightforward: after controlling for AR and cross-lagged effects, the value of a given variable at t explains variance in another variable also at t. Much like in cross-sectional studies, just because one variable is on the right-hand side of the equation does not mean it *causes* the variable on the left-hand side of the equation, although causal claims have been for relations found to exist in data-driven searches made among contemporaneous variables. In any case, modeling contemporaneous relations as directed paths among variables carries with it the assumption that the variables in the system can explain each other.

However, it is also possible that two variables in a system relate to each other due to a third variable that is not in the system. Here, they would be considered exogenously related, where one variable does not explain greater variance than the reverse. The residuals of the variables (after accounting for AR and cross-lagged effects as well as other contemporaneous variables) may covary across time due to a third variable that exists outside of the system, or a common cause. In this case, forcing the model to select a directed relation among the two variables may lead to incorrect inferences. VAR models for this type of relation of residuals.

In Chapter 4, we demonstrated that an SVAR (uSEM) can be transformed into an equivalent VAR where any contemporaneous directed relations among the observed variables surface as both (1) cross-lagged relations and (2) covariances among residuals. Given that this is a transformation, the models are equivalent. Both directed contemporaneous relations among variables and covariances among residuals can be present in hybrid-GIMME. The stepwise search procedures within GIMME can appropriately detect the data-generating relations even when a combination of directed relations among variables and bidirectional covariances among residuals exist (Luo et al., 2022; Molenaar, 2017).

For the hybrid version of GIMME, the **B** (provided in section 6.3.4.1) matrix is identical to that of the uSEM. However, now the covariance matrix of the residuals may include non-zero covariances among the contemporaneous residuals:

$$\Psi^* = \begin{pmatrix} \psi_{11} & & & & & & & & \\ \psi_{21} & \ddots & & & & & & & \\ \vdots & \ddots & \ddots & & & & & & \\ \vdots & & \ddots & \ddots & & & & & \\ \psi_{p1} & \cdots & \cdots & \psi_{p(p-1)} & \ddots & & & & \\ 0 & \cdots & \cdots & \cdots & 0 & \ddots & & & \\ \vdots & \ddots & & & \vdots & \psi_{(p+1)(p)} & & & \\ \vdots & & \ddots & & \vdots & \vdots & \ddots & \ddots & \\ \vdots & & & \ddots & \vdots & \vdots & & \ddots & \ddots \\ 0 & \cdots & \cdots & \cdots & 0 & \psi_{(2p)(p)} & \cdots & \cdots & \psi_{(2p)(2p-1)} & \psi_{(2p)(2p)} \end{pmatrix}_{2p \times 2p}$$

Recall that in regular uSEM, the lower right-hand *pxp* corner is constrained to zero. The space of potential relations is thus expanded. The search procedure is the same as with the original GIMME, except now the elements of the covariance matrix among residuals pertaining to contemporaneous variables are included in the search. Results show that the approach has an excellent recovery of both directed paths among observed variables as well as undirected relations among the residuals (i.e., covariances).

Figure 6.9 depicts results obtained using the same data explored for the GIMME example above but with the hybrid GIMME option utilized. A comparison between Figures 6.8 and 6.9, which depicts the original GIMME results, reveals a number of similarities. Specifically, the contemporaneous relation between Angry and Irritable was recovered as directed rather than as a covariance among residuals in hybrid GIMME much like in the original GIMME results. This suggests that controlling for other variables and the AR effect of Angry on itself, the value of Irritable at a given time point explains variance in Angry for the majority of individuals. A number of individual-level covariance relations among residuals were also found, with the covariance of residuals for Positive and Anhedonia being found the greatest number of times as indicated by the thicker line.

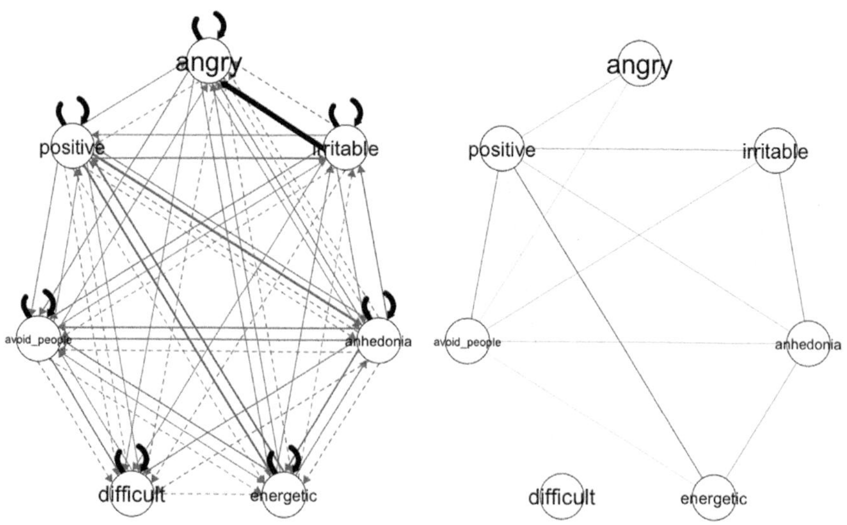

FIGURE 6.9
Hybrid GIMME results for Fisher example. Solid lines represent contemporaneous (lag-0) relations and dashed lines represent lag-1 relations. Line width corresponds to the proportion of individuals with that path. Black lines are group-level paths and were estimated for everyone. Left panel depicts directed relations; right panel depicts bidirectional covariances.

6.8 Conclusions

We presented here first a broad overview of machine learning frameworks supplemented with specific examples that relate to IAV analysis. As with all exploratory methods, the ones presented here must be used with caution. It must be stressed that studies using exploratory methods must explicitly report this, and take great care to validate the results. One of the strongest ways to validate methods is to find if it has a relationship to something outside of the variables included in the exploratory analyses (Aldenderfer & Blashfield, 1984), and that is one approach the current authors strongly encourage. Data-driven searches can also be used for hypothesis generation of causal mechanisms that can be replicated in future confirmatory research or used in clinical practice.

Another important issue is that individual-level exploratory analysis runs the risk of modeling error or fitting a model to noise. The GIMME approaches provide one solution. The core of the GIMME heuristic is looking for relations that exist for the majority of individuals early in the model building. This helps to decrease the level of false positives and missed relations both at the group (sample) and individual levels. At the group level, by looking for consistencies across individuals rather than aggregating the algorithm obtains reliable results that describe relations that likely occur for the sample, and can be generalized to the population from which the data are drawn.

At the individual level, by starting with some known information (here, the group-level paths), we begin the search closer to the true model and, in doing so, improve the recovery of the relations (Gates & Molenaar, 2012).

This chapter utilized VAR, SVAR (uSEM), and hybrid-VAR models. These models quantify contemporaneous relations differently. As discussed in Chapter 5, in uSEM, contemporaneous dependencies are modeled as directed relations among the observed variables. In VAR, they are modeled as bidirectional (correlation) or unidirectional (partial correlation) relations among the residuals. This has important implications for interpretation. When the relations are among the residuals, the underlying assumption is that the two variables are related via a common cause that is outside the variables in the current system. Conversely, when the relations are modeled among the observed variables, the assumption is that the variables that explain variability in each other are within the system of variables provided. Researchers need to think critically about which assumption best matches their data and hypotheses. Alternatively, one can use the hybrid model introduced here.

Many of these developments generalize to other approaches for model building and can be adapted to allow various forms of models. For instance, graphically depicted VAR models can be obtained using a feedforward approach, and uSEM and hybrid uSEM models can be obtained using regularization (Ye, Gates, Henry, & Luo, 2021). Outside of the work presented here, Bayesian approaches for inducing sparsity exist, which is outside the scope of this book which focuses on frequentist methods. The reader interested in Bayesian approaches to VAR modeling is referred to Ahelegbey, Billio, and Casarin (2016) as well as innovations such as those presented in Paci and Consonni (2020).

References

Ahelegbey, D.F., Billio, M., & Casarin, R. (2016). Sparse graphical vector autoregression: A Bayesian approach. *Annals of Economics and Statistics/Annales d'Économie et de Statistique*, (123/124), 333–361.

Aldenderfer, M.S. & Blashfield, R.K. (1984). *Cluster analysis: Quantitative applications in the social sciences.* Beverly Hills, CA: Sage Publication.

Argyriou, A., Evgeniou, T., & Pontil, M. (2007). Multi-task feature learning. *Advances in Neural Information Processing Systems*, 19.

Arizmendi, C., Gates, K., Fredrickson, B., & Wright, A. (2021). Specifying exogeneity and bilinear effects in data-driven model searches. *Behavior Research Methods, 53*(3), 1276–1288.

Beltz, A.M., Gates, K.M., Engels, A.S., Molenaar, P.C., Pulido, C., Turrisi, R., ... & Wilson, S.J. (2013). Changes in alcohol-related brain networks across the first year of college: A prospective pilot study using fMRI effective connectivity mapping. *Addictive Behaviors, 38*(4), 2052–2059.

Beltz, A.M. & Molenaar, P.C. (2016). Dealing with multiple solutions in structural vector autoregressive models. *Multivariate Behavioral Research, 51*(2–3), 357–373.

Brown, T.A. (2006). *Confirmatory factor analysis for applied research*. New York, NY: Gilford Press.

Bollen, K.A. (1989). *Structural equations with latent variables*. New York, NY: Wiley.

Dahlhaus, R. & Eichler, M. (2003). Causality and graphical models in time series analysis. In *Oxford Statistical Science Series* (pp. 115–137).

Duffy, K.A., Fisher, Z.F., Arizmendi, C.A., Molenaar, P.C.M., Hopfinger, J., Cohen, J.R., ... & Gates, K. (2021). Detecting task-dependent functional connectivity in GIMME with person-specific hemodynamic response functions. *Brain Connect, 11*, 418–429.

Engle, R.F. (1984). Wald, likelihood ratio, and Lagrange multiplier tests in econometrics. *Handbook of Econometrics, 2*, 775–826.

Epskamp, S. (2017). graphicalVAR: Graphical VAR for experience sampling data. [Computer software] *R package version 0.2*, https://CRAN.R-project.org/package=graphicalVAR.

Epskamp, S., Waldorp, L.J., Mõttus, R., & Borsboom, D. (2018). The Gaussian graphical model in cross-sectional and time-series data. *Multivariate Behavioral Research, 53* (4), 453–480.

Fisher, A.J. & Boswell, J.F. (2016). Enhancing the personalization of psychotherapy with dynamic assessment and modeling. *Assessment, 23*(4), 496–506.

Friedman J, Hastie T, & Tibshirani R (2010). Regularization paths for generalized linear models via coordinate descent. *Journal of Statistical Software, 33*(1), 1–22. http://www.jstatsoft.org/v33/i01/.

Gates, K.M., Fisher, Z.F., Arizmendi, C., Henry, T.R., Duffy, K.A., & Mucha, P.J. (2019). Assessing the robustness of cluster solutions obtained from sparse count matrices. *Psychological Methods, 24*(6), 675.

Gates, K.M., Fisher, Z.F., & Bollen, K.A. (2020). Latent variable GIMME using model implied instrumental variables (MIIVs). *Psychological Methods, 25*(2), 227.

Gates, K.M., Lane, S.T., Varangis, E., Giovanello, K., & Guiskewicz, K. (2017). Unsupervised classification during time-series model building. *Multivariate Behavioral Research, 52*(2), 129–148.

Gates, K.M. & Molenaar, P.C. (2012). Group search algorithm recovers effective connectivity maps for individuals in homogeneous and heterogeneous samples. *NeuroImage, 63*(1), 310–319.

Goodfellow, I., Bengio, Y., & Courville, A. (2016). *Deep learning*. MIT Press.

Granger, C.W. (1969). Investigating causal relations by econometric models and cross-spectral methods. *Econometrica: Journal of the Econometric Society*, 424–438.

Henry, T. & Gates, K. (2017). Causal search procedures for fMRI: Review and suggestions. *Behaviormetrika, 44*(1), 193–225.

Hillary, F.G., Medaglia, J.D., Gates, K., Molenaar, P.C., Slocomb, J., Peechatka, A., & Good, D.C. (2011). Examining working memory task acquisition in a disrupted neural network. *Brain, 134*(5), 1555–1570.

Lane, S. T., Gates, K., Fisher, Z., Arizmendi, C., Molenaar, P., Hallquist, M., ... & Gates, M.K.M. (2022). Package 'gimme: Group Iterative Multiple Model Estimation' [Computer software]. https://cran.r-project.org/web/packages/gimme/index.html

Lane, S.T., Gates, K.M., Pike, H.K., Beltz, A.M., & Wright, A.G. (2019). Uncovering general, shared, and unique temporal patterns in ambulatory assessment data. *Psychological Methods, 24*(1), 54.

Lauritzen, S.L. (1996). *Graphical models*. Oxford: Clarendon Press.

Luo, L., Fisher, Z.F., Arizmendi, C., Molenaar, P., Beltz, A., & Gates, K.M. (2022). Estimating both directed and undirected contemporaneous relations in time series data using hybrid-group iterative multiple model estimation. *Psychological Methods*.

MacCallum, R.C., Roznowski, M., & Necowitz, L.B. (1992). Model modifications in covariance structure and the problem of capitalization on chance. *Psychological Bulletin, 111*(3), 490–504. PMID: 16250105

Molenaar, P.C. (2017). Equivalent dynamic models. *Multivariate Behavioral Research, 52*(2), 242–258.

Mumford, J.A. & Ramsey, J.D. (2014). Bayesian networks for fMRI: a primer. *Neuroimage, 86*, 573–582.

Ogutu J.O., Schulz-Streeck T., & Piepho H.-P. (2012). Genomic selection using regularized linear Regression models: ridge regression, lasso, elastic net and their extensions. *BMC Proceeding., 6*(2), S10. doi:10.1186/1753-6561-6-S2-S10

Paci, L. & Consonni, G. (2020). Structural learning of contemporaneous dependencies in graphical VAR models. *Computational Statistics & Data Analysis, 144*, 106880.

Quinlan, J.R. (1986). Induction of decision trees. *Machine Learning, 1*(1), 81–106.

Ramsey, J.D., Hanson, S.J., & Glymour, C. (2011). Multi-subject search correctly identifies causal connections and most causal directions in the DCM models of the Smith et al. simulation study. *NeuroImage, 58*(3), 838–848.

Smith, S.M., Miller, K.L., Salimi-Khorshidi, G., Webster, M., Beckmann, C.F., Nichols, T.E., ... & Woolrich, M.W. (2011). Network modelling methods for FMRI. *Neuroimage, 54*(2), 875–891.

Sörbom, D. (1989). Model modification. *Psychometrika, 54*(3), 371–384.

Tibshirani, R. (1996). Regression shrinkage and selection via the lasso. *Journal of the Royal Statistical Society: Series B (Methodological), 58*(1), 267–288.

Wald, A. (1943). Tests of statistical hypotheses concerning several parameters when the number of observations is large. *Transactions of the American Mathematical society, 54*(3), 426–482.

Waldmann, P., Mészáros, G., Gredler, B., Fürst, C., & Sölkner, J. (2013). Evaluation of the lasso and the elastic net in genome-wide association studies. *Frontiers in Genetics*, 4. doi: 10.3389/fgene.2013.00270.

Weigard, A., Lane, S., Gates, K., & Beltz, A. (2021). The influence of autoregressive relation strength and search strategy on directionality recovery in group iterative multiple model estimation. *Psychological Methods.*

Ye, A., Gates, K.M., Henry, T.R., & Luo, L. (2021). Path and directionality discovery in individual dynamic models: A regularized unified structural equation modeling approach for hybrid vector autoregression. *Psychometrika, 86*(2), 404–441.

Zou, H. & Hastie, T. (2005). Regularization and variable selection via the elastic net. *Journal of the Royal Statistical Society: Series B (Statistical Methodology), 67*(2), 301–320. doi: 10.1111/j.1467-9868.2005.00503.x@10.1111/(ISSN)1467-9868.TOP_SERIES_B_RESEARCH.

7

Models of Intra-Individual Variability with Time-Varying Parameters (TVPs)

The methods presented thus far in this book require that the assumption of stationarity be met. That is, the modeled processes are assumed to comprise subcomponents that maintain the same lawful relations over time, with coefficient estimates that are constant across all repeated measurements of the same process. In other words, the modeled processes are expected to show constant means, variances, and covariances over time; and measurement invariance as evidenced through time-invariant factor loadings, intercepts, and related parameters. Chapter 4 further discusses the notion of stability (Lütkepohl, 2005), obtained when stationary systems show values that are bounded within a finite range, as opposed to "explosive" trajectories that show increasingly extreme numerical values approaching positive or negative infinity in the long run. However, instances of nonstationarity abound in everyday life. For example, interpersonal dynamics within dyads can and do change even during a brief episode of interaction as the dyad members undergo different phases and kinds of interactions. Trends (e.g., linear, as addressed in Chapter 4 and Molenaar, de Gooijer, & Schmitz, 1992; or exponential, as investigated in Browne, 1993) constitute another form of nonstationarity that is often dealt with explicitly through the use of models targeting relatively smooth and gradual change processes such as growth curve models (Browne & du Toit, 1991; McArdle & Epstein, 1987; Meredith & Tisak, 1990); or removed from data prior to analysis via time series techniques—a procedure known as detrending (Craigmile, Peruggia, & Van Zandt, 2009). Other well-known examples of nonstationarity have also been noted and replicated time and again in psychomotor experiments. As an example, the Haken, Kelso and Bunz (HKB; 1985) model of interlimb coordination was proposed as a way to represent how continuous changes in the driving frequency of two index fingers' oscillatory motions can transition into sudden shifts from in-phase to out-of-phase (i.e., asymmetrical) motions.

Many methods for detecting and modeling nonstationarity have been proposed over the years. Models with varying coefficients are one possibility that allows parameters in a model to show variations over time or as a function of other known variables such as space, cohort, context, and so on (Fan & Zhang, 2008; Harvey & Phillips, 1982; Hastie & Tibshirani, 1993;

DOI: 10.1201/9780429172649-7

Pagan, 1980). Particularly relevant to the interests of many developmental, social behavioral and health scientists are models with time-varying parameters (TVPs; Molenaar, 1987; Tarvainen et al., 2006; West & Harrison, 1989). The operating premise of such models is that the TVPs change over time at a time scale that is substantially slower than the time scale(s) of the key variables in the system (e.g., the endogenous or dependent latent variables). Imposing time invariance assumptions on systems characterized by TVPs is essentially akin to trying to understand how a system changes without accounting for the fact the system changes *differently* depending on contextual or other factors.

Models with TVPs were referred to as *self-organizing* state-space models (Kitagawa, 1998) because the over-time changes in parameters allow a system to "self-organize" into (possibly) qualitatively distinct forms. In other scientific contexts, researchers have described self-organization as the emergence of organized spatial, temporal, or functional structures or behaviors from random initial conditions, often without clear, specific interference from the outside (Haken, 2006; Rocha, 1998). Freeing up time invariance constraints on modeling parameters provides a mechanism to evaluate evidence for time-related self-organization as testable hypotheses.

This chapter focuses on one important extension of the DFM particularly relevant for the modeling of intra-individual variations—DFM with TVPs. In the explanations that follow, we focus on scenarios in which the parameters vary over time and discuss several possible applications of such models. Formally, we refer to the models discussed in this Chapter as DFM(p,q,l,m,n) with TVPs. As introduced in Chapter 5 of this book, p denotes the number of observed variables, q the number of latent variables, l the maximum number of lags in the measurement model, and m and n are, respectively, the order of the autoregressive and moving average components in the VARMA structure of the structural model. A difference in this chapter is that the DFM is applicable to scenarios where $N \geq 1$. That is, in cases where there are sufficient repeated measurement occasions from an individual to identify all parameters of a DFM, including the subset of coefficients that vary within that individual over time, the model can be fitted separately to each individual's data to yield individual-level estimates. In contrast, when it is tenable to assume that a DFM structure of choice can reasonably capture the patterns of change across a group of individuals, a researcher may gain improvements in estimation accuracy and precision by specifying a group-based DFM composed mostly of parameters that generalize (i.e., are invariant) across individuals, except for a subset of TVPs that may show testable time—as well as individual-varying—dynamics over time. In the remaining of this chapter, we provide step-by-step explications of how such DFMs with TVPs can be used to explore evidence for potential deviations from stationarity, starting with a re-expression of the DFM model for data across $N \geq 1$ individuals.

7.1 The DFM(p,q,l,m,n) across $N \geq 1$ Individuals

The general Dynamic Factor Model, denoted as DFM(p,q,l,m,n), includes concurrent and lagged factor loadings in the measurement model, and VARMA-type relations among a set of latent variables in the structural (also known as dynamic or transition) equation. The DFM(p,q,l,m,n) is reiterated here for convenience as

$$\mathbf{y}_i(t) = \sum_{u=0}^{l} \mathbf{\Lambda}(u)\boldsymbol{\eta}_i(t-u) + \boldsymbol{\epsilon}_i(t) \tag{7.1}$$

$$\boldsymbol{\eta}_i(t) - \sum_{u=1}^{m} \mathbf{\Phi}(u)\boldsymbol{\eta}_i(t-u) = \boldsymbol{\zeta}_i(t) + \sum_{u=1}^{n} \mathbf{\Theta}(u)\boldsymbol{\zeta}_i(t-u) \tag{7.2}$$

where $\mathbf{y}_i(t)$, $t = 1,\ldots,T$, represents the p-variate observed time series for person i, $i = 1,\ldots,N$; $\boldsymbol{\eta}_i(t)$ represents the q-variate latent factor series for person i, and $\boldsymbol{\epsilon}_i(t)$ is a p-variate measurement error series for person i. Note that as the parameter matrices $\mathbf{\Lambda}$, $\mathbf{\Phi}$, and $\mathbf{\Theta}$ do not vary across individuals in the version of the DFM used here, the i subscripts are absent. When this model is fitted to single-subject data with $N = 1$ as in earlier chapters, these parameter estimates are, by nature of the model fitting procedures, individual-specific.

This general form, denoted as DFM(p,q,l,m,n), is distinct from the standard form of the state-space model in its explicit inclusion of the influence of lagged factor loadings on manifest variables in the measurement model. However, as explained in Chapter 5, this model can also be specified as a special case of the state-space model. Allowing for over-time changes in modeling parameters violates one of the fundamental assumptions of models such as DFM(p,q,l,m,n), namely, the stationarity as well as stability assumptions noted earlier.

7.2 The DFM(p,q,l,m,n) with TVPs as a State-Space Model

To fit models with TVPs, it is convenient to re-express the DFM(p,q,l,m,n) in state-space form as introduced in Chapter 5. The state-space model is well suited for handling long multivariate time series, where the length (T) of the time series exceeds the number of subjects, N (i.e., $T \gg N$; Chow et al., 2010; MacCallum and Ashby, 1986; Otter, 1986). As a reminder, to implement the transformation into state-space form, the caveat here is to put latent variables whose lagged versions are needed into an expanded latent variable series, denoted previously as $\boldsymbol{\eta}_{SS}(t)$. Included in the expanded latent variable series, $\boldsymbol{\eta}_{SS,i}(t)$, are lagged versions of $\boldsymbol{\eta}_i(t)$ and $\boldsymbol{\zeta}_i(t)$ up to the desired orders: namely, $\boldsymbol{\eta}_i(t-u)$ for $u = 0, \ldots, \max(l, m-1)$, whichever one is larger; and $\boldsymbol{\zeta}_i(t-u)$, $u = 0, \ldots, n$. Then, we can rewrite Equation (7.1) in an expanded form as

$$\mathbf{y}_i(t) = \boldsymbol{\Lambda}_{SS}\boldsymbol{\eta}_{SS,i}(t) + \boldsymbol{\epsilon}_i(t),\ \boldsymbol{\epsilon}_i(t) \sim N(\mathbf{0}, \boldsymbol{\Theta}) \tag{7.3}$$

Furthermore, with the expanded form of $\boldsymbol{\Phi}_{SS}$, and the expanded process noise vector, $\boldsymbol{\zeta}_{SS,i}(t) = \mathbf{Z}\boldsymbol{\zeta}_i(t)$, the DFM$(p,q,l,m,n)$ model can now be rewritten in state-space form as

$$\boldsymbol{\eta}_{SS,i}(t) = \boldsymbol{\Phi}_{SS}\ \boldsymbol{\eta}_{SS,i}(t-1) + \boldsymbol{\zeta}_{SS,i}(t),\ \boldsymbol{\zeta}_{SS,i}(t)(\mathrm{t}) \sim N(\mathbf{0}, \boldsymbol{\Psi}_{SS}). \tag{7.4}$$

As in the group-based DFM(p,q,l,m,n) described earlier, the state-space model summarized in Equations (7.3)–(7.4) assumes person-invariant parameter matrices such as the factor loading matrix, $\boldsymbol{\Lambda}_{SS}$, and the matrix of auto- and cross-regression coefficients, $\boldsymbol{\Phi}_{SS}$. Person-specific parameter estimates can be obtained by performing model fitting at the individual level, as discussed in Chapters 3–5 for single-subject ($N = 1$) time series data. Throughout this chapter, the terms states and latent variables are used interchangeably.

Several estimation procedures have been proposed in the literature for fitting models with TVPs. We review one approach commonly used in the state-space literature in which the state-space model summarized in Equations (7.3)–(7.4) is expanded to incorporate the TVPs into $\boldsymbol{\eta}_{SS,i}(t)$. That is, the TVPs are now represented as latent variables that are governed by their respective dynamic functions. We refer to the manifest and latent variables that are present in the original model before expansion as *system variables*. The size of the original latent variable vector is still denoted as q. The size of the expanded latent variable following the insertion of TVPs is denoted henceforth as q_{SS}.

To see how TVPs are represented as latent variables, consider the common model featuring associations between a time series of p manifest (i.e., observed) variables and a common factor, $\eta_i(t)$:

$$\mathbf{y}_i(t) = \boldsymbol{\Lambda}\eta_i(t) + \boldsymbol{\epsilon}_i(t), \tag{7.5}$$

which is a special case of the DFM, Cattell's P-technique model described in Chapter 3 (Cattell, Cattell, & Rhymer, 1947; Molenaar & Nesselroade, 2009; Nesselroade & Ford, 1985). The goal of the model is to extract patterns of covariations in multiple manifest variables measured over time, and estimate how they relate to some unmeasured latent construct that also varies across time. This model was initially developed for time series data of a single subject. Later, it was expanded to multiple-subject data assuming homogeneity in patterns of association across individuals.

Equation (7.5) posits invariant (over time) associations among the common factor, $\eta_i(t)$, and the manifest variables in $\mathbf{y}_i(t)$. In other words, longitudinal measurement invariance is explicitly imposed (Meredith, 1993), implying that the parameters in $\boldsymbol{\Lambda}$ do not vary across time. Now suppose such associations, in fact, vary over time as

$$\mathbf{y}_i(t) = \boldsymbol{\Lambda}(t)\eta_i(t) + \boldsymbol{\epsilon}_i(t). \tag{7.6}$$

where some of the values in $\boldsymbol{\Lambda}(t)$ may now change across time. To give a concrete example, suppose that $\mathbf{y}_i(t)$ consists of four manifest variables to identify the common factor, $\eta_i(t)$. In this case, Equation (7.6), which also constitutes the measurement model in a state-space model, now comprises four *nonlinear* equations involving the interaction between two latent variables, $\lambda_{ji}(t)$, and $\eta_i(t), j = 1, \ldots, 4$, as

$$\begin{aligned} y_{1i}(t) &= \lambda_{1i}(t)\eta_i(t) + \epsilon_{1i}(t), y_{2i}(t) = \lambda_{2i}(t)\eta_i(t) + \epsilon_{2i}(t) \\ y_{3i}(t) &= \lambda_{3i}(t)\eta_i(t) + \epsilon_{3i}(t), y_{4i}(t) = \lambda_{4i}(t)\eta_i(t) + \epsilon_{4i}(t) \end{aligned} \tag{7.6a}$$

The latent variable vector, $\boldsymbol{\eta}_{SS,i}(t) = [\eta_i(t)\ \lambda_{1i}(t)\ \lambda_{2i}(t)\ \lambda_{3i}(t)\ \lambda_{4i}(t)]'$, is now expanded to include the common factor as well as four time-varying (TV) factor loadings. Thus, the original vector of latent system variables, $\boldsymbol{\eta}_i(t) = [\eta_i(t)]$ consists only of $q = 1$ latent variable. The expanded latent variable vector, $\boldsymbol{\eta}_{SS,i}(t)$, is of size $q_{SS} \times 1 = 5 \times 1$.

The system depicted in Equation (7.6) violates longitudinal invariance and hence both the stationarity and stability assumptions. At first sight, this seems to violate one of the fundamental premises that enable researchers to study the change in the first place—if the factor loadings vary over time, then there is no basis to claim that the same latent factor, $\eta_i(t)$, is being measured over time. By extrapolation, how can we justify studying changes or intra-individual variability in $\eta_i(t)$ over time when we do not know what it is that $\eta_i(t)$ represents?

Numerous examples exist of measurement tools changing across time to measure individuals. Take, for instance, developmental changes across the lifespan, where psychometric properties (e.g., factor loadings or intercepts) of some items in a questionnaire may and often do change. Items that serve to measure perceptual motor skills in toddlers would be too easy for older children; failure to perform the corresponding tasks might be indicative of other atypical developmental challenges, as opposed to the original construct of interest. This was the reason that motivated researchers to use different measures to assess constructs such as intelligence in children of different ages. As another example, the Wechsler Intelligence Scale for Children (WISC; Wechsler, 2014) has been used as an intelligence test for children between the ages of 6 and 16. For those between 16 and 89 years old, variations of the Wechsler Adult Intelligence Scale are typically used (Drozdick et al., 2012; Wechsler, 1981). The extent to which intelligence-related processes can explain the covariations among items and subtests may vary even within the age bracket targeted by the tests. Still, the psychometric properties should be changing slowly enough that homogeneity and constancy can be expected within shorter windows of time.

The same logic can be applied to studies of individuals across shorter periods of time where the time points are densely gathered (i.e., time series analysis). Studying change in the presence of TVPs is plausible when changes in the TVPs occur at much slower time scales than those associated with other variables in the model (e.g., $\mathbf{y}_i(t)$ and $\eta_i(t)$). As such, the trajectories of the TVPs and relatedly, those of the system variables, can be uniquely identified within local segments of the data. This key assumption helps to render a model with TVPs identified, allowing for estimation of the TVPs and other time-invariant parameters. In this sense, such changes are indeed common across a variety of "real-world" settings.

As in all state-space models, dynamic functions need to be specified for each of the latent variables in $\boldsymbol{\eta}_{SS,i}(t)$. Any of the functions used in this book to describe intra-individual variations may be used as the dynamic function for the common factor, $\eta_i(t)$. For instance, $\eta_i(t)$ may follow an AR(1) process as

$$\eta_i(t) = \phi_1\eta_i(t-1) + \zeta_i(t), \tag{7.7a}$$

or even a function dictating no change as $\eta_i(t) = \eta_i(t-1)$.

What shapes or functions might be appropriate to represent changes in TVPs? The random walk (RW) model, one of the models illustrated in this chapter, is a popular candidate function (Molenaar & Newell, 2003). This model specifies that $\lambda_{ji}(t)$ changes as

$$\lambda_{ji}(t) = \lambda_{ji}(t-1) + \zeta_{\lambda_{ji}}(t), \tag{7.7b}$$

which relates the value of the jth factor loading at time t to its value at time $t - 1$, and $\zeta_{\lambda_{ji}}(t)$ represents the process noise for the jth factor loading. The RW model is a popular function for representing TVPs because it includes the special case of a time-invariant ("no change") model when the process noise variance, Var[$\zeta_{\lambda_{ji}}(t)$], is equal to zero for all js.

Another straightforward way to represent changes in TVPs is to allow for changes in parameter values as a function of developmental stages, such as

$$\lambda_{ji}(t) = \lambda_{j0} + \lambda_{j1} isAgeGE16_i(t), \tag{7.7c}$$

in which $isAgeGE16_i(t)$ is a dummy-coded variable (=1 when individual i is 16 years or older at time t, and 0 otherwise), λ_{j0} is the jth factor loading value for those younger than 16 years old, and λ_{j1}captures the deviation in value of the jth factor loading when individual i is 16 years old and beyond. Similarly, other indicators of experimental conditions can be incorporated into the equations for TVPs that capture across-condition variations in modeling parameters, as was done by Chen et al. (2020) to capture over-time variations in the AR parameters characterizing mother-infant dyads' head movements in the context of an experimental procedure known as the Face-to-Face/Still-Face (FFSF) procedure. During the FFSF procedure, mother-infant dyads were asked to engage in an episode of interactive play (Face-to-Face; FF), followed by a Still-Face (SF) episode during which the mothers were instructed to show disengagement by maintaining a still face to the infants, and finally, a Reunion (RE) episode during which the mothers resumed playing with the infants. Chen et al. found that infants' AR levels in head movements decreased from the FF to the SF episode, and remained low during RE, suggesting that the infants were making less consistent movements in SF and RE as compared to in FF. Mothers' AR levels in head movements also showed patterns of variations that are consonant with the nature of the FFSF manipulation, and evidence of moderating effects of mother smiling and attachment strengths with the infants.

When no explicit information is available to inform the timing and nature of changes in the TVPs, a nonparametric model or related variations that are deemed flexible enough to capture a variety of different change trajectories is used to approximate the changes in the TVPs. Molenaar (1994) considered a one-factor DFM model in which a first-order autoregressive [AR(1)] function is used to describe the changes in $\eta_i(t)$, and polynomial functions of time were used to represent the dynamics of the TVPs, including the AR(1) and the factor loading parameters. Similar polynomial functions were used by Oud and Jansen (2000) to allow for TVPs in the context of fitting linear stochastic differential equation models within the structural equation modeling framework. Many other variations are possible, resulting, for instance, in models such as the local level model (Durbin & Koopman, 2001), the local linear trend model (Harvey, 2001), TV autoregressive moving average (ARMA) model (Tarvainen et al., 2006; Weiss, 1985), and stochastic regression model (Pagan, 1980). We will discuss and provide sample code to implement some of these relatively flexible functions for TVPs in this chapter.

The main complexity that arises with the incorporation of TVPs as latent variables with their own dynamic functions is that the state-space model (e.g., Equation 7.6) becomes nonlinear. For instance, Equation (7.6) now involves the multiplication between two sets of latent variables, $\boldsymbol{\Lambda}(t)$ and $\eta_i(t)$. Thus, methods that explicitly handle the nonlinearity are needed. In the following sections, we describe how this approach and the associated estimation procedures are implemented, followed by illustrations of some special cases of models with TVPs using an R package, Dynamic Modeling in R (dynr; Ou, Hunter, & Chow, 2019).

7.3 Nonlinear State-Space Model Estimation Methods

We have indicated that the incorporation of TVPs in a model as additional latent variables into an originally linear state-space model renders the model nonlinear. That is, the model then includes nonlinear functions involving the latent variables. To allow TVPs to be handled in this manner, one possibility is to first modify Equation (7.6) so that it allows for nonlinearity in the latent variables, within which Equation (7.6) can be treated as a special case. Adding a vector of fixed TV covariates, $x_i(t)$, to the model (e.g., dummy or other coding variables to identify different age segments or experimental conditions), this nonlinear model can be expressed as

$$\begin{gathered}\boldsymbol{\eta}_{SS,i}(t)=\boldsymbol{g}\left(\boldsymbol{\eta}_{SS,i}(t-1),\,\boldsymbol{\vartheta},\mathbf{x}_i(t)\right)+\boldsymbol{\zeta}_{SS,i}(t),\,\boldsymbol{\eta}_{SS,i}(1)\sim N(\mathbf{a},\boldsymbol{\Sigma}_0),\\ \mathbf{y}_i(t)=\boldsymbol{h}\left(\boldsymbol{\eta}_{SS,i}(t),\,\boldsymbol{\vartheta},\mathbf{x}_i(t)\right)+\boldsymbol{\epsilon}_i(t),\\ \boldsymbol{\zeta}_{SS,i}(t)\sim N(0,\boldsymbol{\Psi}_{SS}),\,\boldsymbol{\epsilon}_i(t)\sim N(0,\boldsymbol{\Theta}),\,t=1,\ldots,T;i=1,\ldots N,\end{gathered} \tag{7.8}$$

where $\boldsymbol{\eta}_{SS,i}(t)$ is the expanded latent variable vector with TVPs and terms such as $\mathbf{x}_i(t)$, $\boldsymbol{\zeta}_{SS,i}(t)$, $\boldsymbol{\epsilon}_i(t)$, $\boldsymbol{\Psi}_{SS}$, and $\boldsymbol{\Theta}$ are as defined earlier. $\boldsymbol{g}\left(\boldsymbol{\eta}_{SS,i}(t-1),\,\boldsymbol{\vartheta},\mathbf{x}_i(t)\right)$ is a set of differentiable linear or nonlinear regression functions describing the transition of $\boldsymbol{\eta}_{SS,i}(t)$ from time $t-1$ to time t (e.g., Equations 7.7a–7.7b); $\boldsymbol{h}\left(\boldsymbol{\eta}_{SS,i}(t),\,\boldsymbol{\vartheta},\mathbf{x}_i(t)\right)$ is a set of differentiable linear or nonlinear measurement functions specifying the relations between the latent variables and the manifest variables (e.g., Equation 7.6a). $\boldsymbol{\vartheta}$ is a vector of time-invariant parameters that appear in the model. In the common factor model with TV factor loadings defined in Equations (7.6)–(7.7), for example, these time-invariant parameters include the unique (or "measurement error") variances in $\boldsymbol{\Theta}$, any variance-covariance parameters for the process noises in $\boldsymbol{\zeta}_{SS,i}(t)$, and parameters that appear in the initial condition means and covariance matrix, $\mathbf{a}$ and $\boldsymbol{\Sigma}_0$, respectively. In the context of Equation (7.7c), λ_{j0} and λ_{j1} are also examples of time-invariant parameters. $\boldsymbol{\eta}_{SS,i}(1)$ contains the values of the expanded latent variable vector at time 1, or in other words, the initial conditions of the latent variables from which subsequent over-time iterations of $\boldsymbol{\eta}_{SS,i}(t)$ are based.

7.3.1 Estimation Procedures

Nonlinear state-space models share the same merits as linear state-space models in terms of their flexibility in handling time series of arbitrary lengths. However, explicit procedures are needed to handle the nonlinearities in Equation (7.8). Here, we describe a nonlinear Kalman filtering approach that provides one of the simplest and most commonly used nonlinear counterparts to the linear Kalman filter approach described in Chapter 5. This nonlinear approach uses the extended Kalman filter (EKF) and the corresponding extended Kalman smoother (EKS) for latent variable estimation purposes (Gelb, 1974; Molenaar & Newell, 2003). That is, once a model has been expressed in the nonlinear state-space form shown in Equation (7.8), the EKF and the related EKS can be used to derive longitudinal factor or latent variable scores at each time point (i.e., estimates of $\boldsymbol{\eta}_i^*(t)$ conditional on the data). By-products from running the EKF can, in turn, be substituted into a raw data log-likelihood function to be optimized to yield (pseudo-) maximum likelihood estimates (MLEs) for the time-invariant parameters in $\boldsymbol{\vartheta}$.

7.3.1.1 *The Extended Kalman Filter (EKF) and the Extended Kalman Smoother (EKS)*

Suppose the data set $\mathbf{Y}_i(t) = [\mathbf{y}_i^T(1), \mathbf{y}_i^T(2), \ldots, \mathbf{y}_i^T(t)]^T$ denotes the p-variate time series available from person i from time 1 to T. The EKF (Anderson & Moore, 1979) is designed to estimate a subclass of nonlinear state-space models with additive, Gaussian process and measurement noises. Based on the nonlinear state-space model in Equation (7.8), the EKF can be used to derive conditional latent variable score estimates based on manifest observations up to time t, $\boldsymbol{\eta}_{SS,i}(t\,|\,t)$, $\triangleq$ (i.e., defined as) $E[\boldsymbol{\eta}_{SS,i}(t)\,|\,\mathbf{Y}_i(t)]$; and the associated covariance matrix, $\mathbf{P}_i(t\,|\,t) \triangleq \mathrm{Cov}[\boldsymbol{\eta}_{SS,i}(t)\,|\,\mathbf{Y}_i(t)]$. One complete cycle of the EKF involves the successive implementation of a *Prediction Stage* and an *Update Stage*, as outlined below, for $t = 1, \ldots, T$ and $i = 1, \ldots, N$.

Prediction Stage. Within each cycle of the EKF, the prediction stage serves to provide model-implied mean (expected) and covariance values of the latent variables one time step ahead (at time t) based on estimates computed thus far using the observed information from time 1 through t – 1. That is, for the mth iteration cycle, the values of the parameters in $\boldsymbol{\vartheta}$ are first set to some known values. For instance, $\boldsymbol{\vartheta}^m$ may be set to the modeler-supplied starting values, $\boldsymbol{\vartheta}^0$, or the parameter estimates at the mth iteration cycle, $\hat{\boldsymbol{\vartheta}}^m$), the latent variable scores and their associated covariance matrix conditional on $t = 0$ are first set to $\mathbf{a}$ and $\boldsymbol{\Sigma}_0$, respectively. For $t > 1$, these values are predicted from time $t - 1$ to t using the updated state and covariance estimates (to be covered below) from t – 1, $\boldsymbol{\eta}_{SS,i}(t-1\,|\,t-1)$ and $\mathbf{P}_{SS,i}(t-1\,|\,t-1)$, respectively, as

$$\begin{aligned}\boldsymbol{\eta}_{SS,i}(t\,|\,t-1) &\triangleq E\left[\boldsymbol{\eta}_{SS,i}(t)\,|\,\mathbf{Y}_i(t-1)\right] = g(\boldsymbol{\eta}_{SS,i}(t-1\,|\,t-1), \boldsymbol{\vartheta}^m, \mathbf{x}_i(t))\\ \mathbf{P}_{SS,i}(t\,|\,t-1) &\triangleq \mathrm{Cov}\left[\boldsymbol{\eta}_{SS,i}(t)\,|\,\mathbf{Y}_i(t-1)\right] = \mathbf{G}_i(t)\mathbf{P}_{SS,i}(t-1\,|\,t-1)\mathbf{G}_i(t)' + \boldsymbol{\Psi}_{SS},\end{aligned} \tag{7.9}$$

in which $\mathbf{G}$ is the Jacobian matrix containing differentiations of the right-hand side of each of the dynamic functions (e.g., Equations 7.7a–7.7b) with respect to each of the latent variables in $\boldsymbol{\eta}_{SS,i}(t-1)$, evaluated at the values of updated state estimates from the previous time point, $\boldsymbol{\eta}_{SS,i}(t-1\,|\,t-1)$, namely,

$$\mathbf{G}_i(t) = \frac{\partial g\left(\boldsymbol{\eta}_{SS,i}(t-1), \boldsymbol{\vartheta}^m, \mathbf{x}_i(t)\right)}{\partial \boldsymbol{\eta}_{SS,i}(t)}\Big|_{\boldsymbol{\eta}_{SS,i}(t-1|t-1)} \tag{7.9a}$$

where the jth row and kth column of $\mathbf{G}_i(t)$ contain the partial derivative of the jth dynamic function with respect to the kth latent variable, evaluated at subject i's updated latent variable estimates from time $t-1$, $\boldsymbol{\eta}_{SS,i}(t-1\,|\,t-1)$.

The subject index in $\mathbf{G}_i(t)$ is used to indicate that the associated Jacobian matrices have different numerical values because they are evaluated at each person's current state estimates.

Predicted observations are then obtained as

$$\mathbf{y}_i(t|t-1) = \mathbf{h}\left(\boldsymbol{\eta}_{SS,i}(t|t-1), \boldsymbol{\vartheta}^m, \mathbf{x}_i(t)\right). \tag{7.9b}$$

These predicted state and observed variable estimates are computed using state estimates updated using observations up to time $t-1$. They are then passed on to the update stage to complete the EKF computations for time t.

Update Stage. The update stage involves updating or fine-tuning of the state estimates and their corresponding covariance matrix by incorporating the manifest observations available at time t. In this stage, the latent variable estimates and their associated covariance matrix are updated to obtain $\boldsymbol{\eta}_{SS,i}(t|t), \triangleq E[\boldsymbol{\eta}_{SS,i}(t)|\mathbf{Y}_i(t)]$ and the associated covariance matrix, $\mathbf{P}_i(t|t) \triangleq \text{Cov}\left[\boldsymbol{\eta}_{SS,i}(t)|\mathbf{y}_i(t)\right]$ as

$$\begin{aligned}
\boldsymbol{\eta}_{SS,i}(t|t) &= \boldsymbol{\eta}_{SS,i}(t|t-1) + \mathbf{K}_i(t)\mathbf{v}_i(t), \\
\mathbf{P}_{SS,i}(t|t) &= \mathbf{P}_{SS,i}(t|t-1) - \mathbf{K}_i(t)\mathbf{H}_i(t)\mathbf{P}_{SS,i}(t|t-1), \\
\mathbf{v}_i(t) &= \mathbf{y}_i(t) - \mathbf{y}_i(t|t-1), \\
\mathbf{F}_i(t) &= \mathbf{H}_i(t)\mathbf{P}_{SS,i}(t|t-1)\mathbf{H}_i(t)' + \boldsymbol{\Theta}, \\
\mathbf{K}_i(t) &= \mathbf{P}_{SS,i}(t|t-1)\mathbf{H}_{SS,i}(t)'\mathbf{F}_i^{-1}(t),
\end{aligned} \tag{7.10}$$

where $\mathbf{K}_i(t)$ is referred to as the Kalman gain matrix, $v_i(t)$ is the vector of one-step-ahead prediction errors as defined in Chapter 5 that carries the new information available at time t that is not accounted for by the prediction based on observations up to time $t-1$. $\mathbf{H}_i(t) = \partial h\left(\boldsymbol{\eta}_{SS,i}(t), \boldsymbol{\vartheta}^m, \mathbf{x}_i(t)\right) \big/ \partial \boldsymbol{\eta}_{SS,i}(t) \big|_{\boldsymbol{\eta}_{SS,i}(t|t-1)}$ is a Jacobian matrix containing differentiations of each of the measurement functions (e.g., Equation 7.6a) with respect to the latent variables at time t. In particular, the jth column and kth row of $\mathbf{H}_i(t)$ carry the partial derivative of the jth measurement function with respect to the kth latent variable at time t, evaluated at subject i's predicted latent variable estimates at time t, $\eta_{SS,i}(t|t-1)$.

The EKF can be understood as a direct adaptation of the linear KF by replacing the nonlinear dynamic and measurement functions with Taylor series expansion, namely, polynomial functions involving derivatives of the nonlinear functions. The EKF algorithm summarized in Equations (7.9)–(7.10) is a first-order EKF algorithm that retains only the linear terms (i.e., only the first derivatives of the possibly nonlinear dynamic and measurement functions) in its approximation to yield, for each t, the latent variable score estimates and their associated covariance matrix, $\boldsymbol{\eta}_{SS,i}(t|t)$ and $\mathbf{P}_{SS,i}(t|t)$ using only information up to time t, but not information from the entire time series (i.e., up to time T).

Extended Kalman Smoother (EKS). Since not all data are used, a smoother is often applied which takes into account data that are available beyond time t. The fixed interval smoother (Anderson & Moore, 1979) is one possible Kalman smoother, whose function is to derive conditional expected values of the latent variables at any particular time point using all available data, namely, $E[\boldsymbol{\eta}_{SS,i}(t) \mid \mathbf{y}_i(T)]$, denoted herein as $\boldsymbol{\eta}_{SS,i}(t \mid T)$, and the associated covariance matrix, $\mathbf{P}_{SS,i}(t \mid T)$. One version of the smoother (Shumway & Stoffer, 2000), denoted herein as the EKS, can be implemented by performing a backward recursion for $t = T, \ldots, 1$ as

$$\begin{aligned} \boldsymbol{\eta}_{SS,i}(t \mid T) &= \boldsymbol{\eta}_{SS,i}(t \mid T) + \mathbf{J}_i(t)[\boldsymbol{\eta}_{SS,i}(t+1 \mid T) - \boldsymbol{\eta}_{SS,i}(t+1 \mid t)], \\ \mathbf{P}_{SS,i}(t \mid T) &= \mathbf{P}_{SS,i}(t \mid T) + \mathbf{J}_i(t)[\mathbf{P}_{SS,i}(t+1 \mid T) - \mathbf{P}_{SS,i}(t+1 \mid t)]\mathbf{J}_i(t)'. \end{aligned} \tag{7.11}$$

where $\mathbf{J}_i(t) = \mathbf{P}_{SS,i}(t \mid T)\mathbf{G}_i'(t+1)\mathbf{P}_{SS,i}^{-1}(t+1 \mid t)$. In sum, by using all the observations from each individual for state estimation purposes, the EKS typically provides improvements over the EKF by increasing the accuracy (i.e., in reducing average errors) as well as precision (in terms of reducing the variances) of the state estimates, particularly at earlier time points when the EKF has limited observations for estimation purposes. Some exceptions do exist, such as when a system process is completely deterministic (no process noises are present) and the initial conditions are known perfectly.

Other State Estimation Algorithms. The EKF and EKS discussed thus far are just a limited subset of the available algorithms for state estimation purposes. The algorithms implemented in the *dynr* package used for the illustrations in this chapter parallel the EKF and EKS for the discrete-time models considered in this chapter, wherein the data are equally spaced in time, and successive time steps can be conceived discretely as integer multiples of the first time point (i.e., $t = 1, 2, 3, \ldots, T$). For continuous-time models targeting data that may be irregularly spaced, *dynr* uses the continuous-discrete time EKF (CDEKF). In the CDEKF, one particular kind of numerical solvers, the fourth-order Runge–Kutta method, is used to predict numerical values of the latent variables under a continuous-time model of interest—generally appearing in a form referred to as differential equations—in the prediction stage. These predicted state estimates are then refined with measurements observed at discrete time points in the update stage. Interested readers are referred to Ou et al. (2019) and the references therein (see also Bar-Shalom, Li, & Kirubarajan, 2001) for details. For state estimation purposes involving nonlinear dynamic and measurement functions and non-Gaussian data, we refer the readers to some of the ongoing work on particle and Monte Carlo filtering methods (Doucet, de Freitas, & Gordon, 2001; Durbin & Koopman, 2001).

7.3.1.2 Parameter Estimation

The one-step-ahead prediction errors, $v_i(t)$, have known parametric distribution under the assumptions that (1) the measurement functions in

Equation (7.8) are linear conditional on (i.e., under fixed values of) the latent states; and (2) the process noise, measurement noise, and initial latent variable vector, $\eta_{SS,i}(1)$, are Gaussian distributed (Chow et al., 2011; Chow, Ferrer, & Nesselroade, 2007). When these assumptions are met, a log-likelihood function, termed the Prediction Error Decomposition (PED) function (Schweppe, 1965), can be written as a function of $v_i(t)$ and the associated covariance matrix, $\mathbf{F}_i(t)$, yielding (Caines & Rissanen, 1974; Harvey, 2001; Schweppe, 1965):

$$\log L(\theta) = \frac{1}{2}\sum_{i=1}^{n}\sum_{t=1}^{T} - p_{it}\log(2\pi) - \log\left|\mathbf{F}_i(t)\right| - \mathbf{v}_i'(t)\mathbf{F}_i(t)^{-1}\mathbf{v}_i(t), \tag{7.12}$$

where $|.|$ denotes the determinant of a matrix; p_{it} is the number of complete manifest variables at time t for person i, made person-specific to accommodate person-specific missingness patterns and allow the use of all available non-missing observations for computation of the log-likelihood function at each time point, as similar to standard full-information maximum likelihood (FIML) estimation approaches.

Thus, each cycle of execution of the EKF (or other related filtering or state estimation algorithms) at a particular iteration m yields all the by-products necessary (i.e., $v_i(t)$ and $\mathbf{F}_i(t)$ for $t = 1, \ldots, T$ and $i = 1, \ldots, N$) to compute the log-likelihood value at the set of parameter values, $\boldsymbol{\vartheta}^m$. Repeated calls to the EKF over multiple iterations to find the values of $\boldsymbol{\vartheta}$ that maximize the PED function then yield (pseudo-)MLEs of these parameters. Standard errors of all the time-invariant parameters can be obtained by taking the square root of the diagonal elements of $\mathbf{I}^{-1}$, where $\mathbf{I}$ is the observed information (i.e., negative numerical Hessian) matrix at the maximum of the PED function shown in Equation (7.12). We refer to these estimates generally as pseudo MLEs because the distribution of the prediction errors may depart from the Gaussian form assumed in Equation (7.12) in many scenarios, such as when the nonlinearities in the dynamic and measurement functions are not adequately approximated using the state estimation approach of choice, and when Gaussian assumptions for the process noises and initial conditions are untenable. Other modified recursions to improve the numerical stability of the optimization process have been proposed (Ansley & Kohn, 1985; De Jong, 1991; Koopman, 1997) but are beyond the scope of this chapter.

The log-likelihood function in (7.12) can, in turn, be used to compute information criterion measures such as the Akaike information criterion (AIC) and Bayesian information criterion (BIC) as

$$\begin{aligned} \text{AIC} &= -2\log L(\theta) + 2r^* \\ \text{BIC} &= -2\log L(\theta) + r^*\log\left(\sum_{i}^{n} T_i\right), \end{aligned} \tag{7.13}$$

where r^* is the number of parameters in a model and $\Sigma_i^n T_i$ denotes the total number of time points across all individuals. Lower algebraic values of AIC and BIC indicate better fit, including AIC and BIC values that are negative.

In sum, the EKF and EKS can be used to estimate the latent variable scores—including TVPs that are now formulated as latent variables—in models that are special cases of Equation (7.8). Under the additional restriction of linear measurement functions conditional on the state estimates, the log-likelihood function shown in Equation (7.12) can further be optimized to yield estimates of all other time-invariant parameters. Information criterion measures can also be constructed using this log-likelihood function for model comparison purposes.

7.4 Observability and Controllability Conditions in TVPs

To ensure that the dynamics of the system depicted in Equations (7.5)–(7.6) can be recovered and estimated using the available observed data, it is important to verify whether the proposed system is indeed "observable". If a system is observable, then the values of the system at any time can be fully and uniquely determined from observed measurements over a finite time interval. The system in Equation (7.8) is observable if the observability matrix, defined as

$$\begin{bmatrix} \mathbf{H}(t) & \mathbf{G}_i(t)\mathbf{H}(t) & \ldots & (\mathbf{G}_i)^{q_{SS}-1}(t)\mathbf{H}(t) \end{bmatrix}, \tag{7.14}$$

is of full rank, or in other words, has a rank equal to the number of expanded latent variables (i.e., the size of $\eta_{SS,i}(t)$). In Equation (7.10), q_{SS} is the size of $\eta_{SS,i}(t)$, the expanded vector of latent variables. Proofs similar to those proposed by Ou (2018) for checking the observability of state-space models when random effects are included as additional latent variables can be adapted. In our case, the number of permissible TVPs for a state-space model with TVPs to still be observable is restricted by p, the number of manifest variables, and q, the number of latent system variables. If the TVPs appear exclusively in the measurement model, and the factor loading matrix follows a simple structure (as considered, e.g., by Molenaar et al., 1992), then p TVPs in the measurement functions may be incorporated. In contrast, if the TVPs appear exclusively in the dynamic model, then q TVPs in the dynamic functions may be accommodated. Other scenarios in which the TVPs are present in both the dynamic and measurement functions are also possible. The online supplement contains an R script with code and examples for checking the observability of selected state-space models in the online supplementary materials.

Another related property often discussed in the control engineering literature is the controllability property. A system is said to be controllable if it can be driven to assume a particular set of values through manipulation of elements in the vector of exogenous variables, $\boldsymbol{x}_i(t)$. For the purposes of this particular chapter, we are less interested in the issue of controllability than that of observability. In Chapter 8, we return to the issue of controllability.

7.5 Possible Functions for Representing Changes in the TVPs

In most applications, TVPs are not entities that may be directly observed in one's data. However, the state-space approach discussed in this chapter requires that a dynamic function be specified for each of the TVPs as other latent system variables. Deciding on what change functions to use for TVPs is not always easy or natural. Thus, many applications involving TVPs use some nonparametric—namely, model-free—or semiparametric functions to allow for enough flexibility and represent the over-time variations in the TVPs. We review a relatively flexible model known as the generalized RW model for TVPs, proposed by Jakeman and Young (1979, 1984) that subsumes several popular models of TVPs, including the RW and time-invariant models discussed earlier.

We denote person i's kth TVP herein as $\vartheta_{ik}(t)$ to highlight the person- and time-varying nature of such parameters, consistent with our notation for other latent system variables in a state-space model. In the generalized RW model, person i's kth TVP, $\vartheta_{ik}(t)$, is portrayed to vary over time as

$$\begin{aligned} \vartheta_{ik}(t) &= \alpha\vartheta_{ik}(t-1)+\beta\Delta_{ik}(t-1)+\delta\zeta_{\vartheta,ik}(t), \\ \Delta_{ik}(t) &= \gamma\Delta_{ik}(t-1)+\zeta_{\Delta,ik}(t), \end{aligned} \tag{7.15}$$

where the function consists of two latent variables: a local level at time $t-1$, denoted as $\vartheta_{ik}(t-1)$; and a local slope (or local deviation), denoted as $\Delta_{ik}(t-1)$. Each of them has a process noise component: $\zeta_{\vartheta,ik}(t)$ for the local level and $\zeta_{\Delta,ik}(t)$ for the local slope. The local level captures the value of a TVP at each time point, particularly as carried forward from time $t-1$ to time t. The local slope allows for additional deviations in the value of the TVP—consistently positive values of the local slope would give rise to monotonic (i.e., no reversion in change direction) increase or growth in the values of the TVP; consistently negative values of local slope yield monotonic decline in the TVP. In contrast, a mix of positive and negative values yields a trajectory interspersed with periods of ebb and flow in values. Functionally, the local level is often used to instill some continuity or regularity in the values of a TVP over time, whereas the local slope component allows for greater magnitudes of increases and decreases. All of the remaining unknowns are either constants that are fixed at known values to yield specialized functions, or are freely estimated

TABLE 7.1

Different Special Cases of the Generalized RW Model Given in Equation (7.15)

Model	Parameter Constraints	Characteristics
Time-invariant	$\alpha = 1; \delta = \beta = \gamma = 0$	No changes
RW	$\alpha = \delta = 1; \beta = \gamma = 0$	Moving mean that deviates from 0 or IC in a finite window; relatively "choppy" trajectory
Autoregressive	$0 < \alpha < 1; \beta = \gamma = 0; \delta = 1$	Return to 0 in the absence of process noise
IRW	$\alpha = \beta = \gamma = 1; \delta = 0$	Relatively prominent but smooth trend
Smoothed IRW	$0 < \alpha < 1; \beta = \gamma = 1;$ $\delta = 0$	Smooth trajectory that tends toward some equilibrium value except for some local deviations
Local linear trend	$\alpha = \beta = \gamma = \delta = 1;$	Prominent trend that is relatively choppy
Damped linear trend	$\alpha = \beta = \delta = 1; 0 < \gamma < 1$	Local deviations show damping over time; relatively choppy trajectory

Note: RW = random walk; IRW = integrated random walk =; IC = initial condition.

parameters. These special cases and their respective characteristics are summarized in Table 7.1. We will discuss each of them in turn.

The first special case of Equation (7.15) is a time-invariant model in which no changes are observed in the modeled parameter, obtained when $\alpha = 1$; $\delta = \beta = \gamma = 0$. The latter two constraints also imply that the process noise variances for the local level and slope are fixed as zero. If the inclusion of these constraints does not lead to substantial changes in model fit compared to more complex variations with fewer constraints, then one may argue that the parameter of interest may be more parsimoniously regarded as invariant over time. As discussed earlier, this can serve as a starting point to explore the need to allow some of the parameters in a model as TVPs. Improved inferential approaches have also been proposed over the years to enable more accurate assessments of the importance to allow particular parameters as TVPs (Fan & Zhang, 2000; Zhao & Zhao, 2020).

If $\alpha = \delta = 1$ and $\beta = \gamma = 0$, Equation (7.15) yields the RW model discussed earlier:

$$\vartheta_{ik}(t) = \vartheta_{ik}(t-1) + \zeta_{\vartheta,ik}(t). \tag{7.16}$$

This model is a commonly adopted model for TVPs due to its close parallel in form to the time-invariant model, except for the presence of process noise that allows the modeled trajectory to show deviations from its initial condition (IC; starting value). Because of the effect of direct process noise on the local level, $\vartheta_{ik}(t)$, the resultant trajectory tends to appear "choppy" —i.e., showing fine-grained fluctuations—over time.

Under similar constraints as the RW model, but with the requirement that α be less than 1.0 in absolute value, we have the (stationary or stable) autoregressive model:

$$\vartheta_{ik}(t) = \alpha\vartheta_{ik}(t-1) + \zeta_{\vartheta,ik}(t) \tag{7.17}$$

This model yields trajectory that fluctuates around 0, with the tendency to approach and stay at 0 in the absence of any process noise.

A few other variations are described in Table 7.1 and examples of trajectories generated using these functions are plotted in Figure 7.1. Generally, these special cases would yield relatively choppy-looking trajectories when process noises affect the TVPs' local levels directly. Otherwise, if there is no process noise or process noise only affecting the local slopes but not the local levels of the parameter estimates across time, the resultant trajectories would appear as smooth trajectories. Models that incorporate a local slope component tend to yield more prominent trends or movements away from the

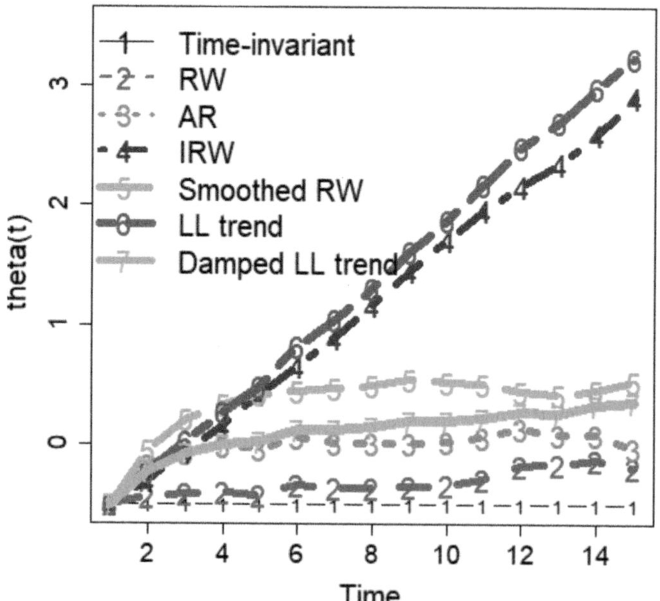

FIGURE 7.1
Plots of trajectories generated using different special cases of the generalized random walk model. RW = random walk; AR = autoregressive; IRW = integrated random walk; LL = local linear.

parameter's IC. Of these models, the integrated random walk (IRW) model, the local linear trend model, and the damped local linear trend model are variations that are more appropriate for capturing prominent trends with large magnitudes of changes.

The smoothed IRW (SIRW) model has identical constraints to the IRW model except that the parameter α is required to be positive and less than 1.0. This model is similar to the autoregressive model with no process noise in that the constraint on α (<1 in absolute value) helps impose smoothing effects. Thus, any large previous deviations would diminish or get "smoothed out" over time. Unlike the autoregressive model, the SIRW explicitly allows process noise only in the local slope but not the local level, and allows for the incorporation of a local slope component. As such, the trajectories generated using the SIRW model tend to be relatively smooth as they "hover around" some equilibrium values (e.g., 0 in Figure 7.1) determined jointly by values of the initial conditions of $\theta_{ik}(t)$, $\Delta_{ik}(t)$, and α.

The local linear trend model is another interesting special case because it includes, as another special case, the linear growth curve model when $\delta = 0$ and $\mathrm{Var}(\zeta_{\Delta,i,k}) = 0$. Because of the incorporation of damping in the local deviations, the damped local linear trend model shows less drastic changes compared to both the IRW model and the local linear trend model. The resultant trajectory is somewhat similar to that obtained under the RW model, however. Thus, the choice of which of these variations of the generalized RW model to use depends largely on the researcher's preconceived notion of the nature of the changes characterizing the TVPs.

7.6 Illustrative Examples

7.6.1 DFM Model with Time-Varying Set-Point

We proceed with a simulation example designed to mirror the multi-time-scale dynamics seen in learning data, in which individuals show overall improvements in learning on a novel task following a sigmoid-shaped learning trajectory, with some intra-individual variations (occasion-to-occasion fluctuations) in improvements over time. The data consist of dynamics from an overall, more-or-less systematic learning trend, with fluctuations around the systematic learning trend. To generate the data, we used a DFM(6,2,0,1,0) model in which six manifest variables were used to identify two latent factors (each identified by three manifest indicators).

The latent factors followed a vector autoregressive relation of order one (i.e., $m = 1$). In addition, we specified a common trend for all the manifest indicators that served to identify the latent factors. This "common trend" may be thought of as a series of gradual intra-individual (e.g., developmental) changes that unfold in a manner similar to how individuals' consistent growth in general (e.g., fluid) intelligence from childhood to adulthood affects multiple domains of ability, including their verbal and arithmetic abilities. We generated simulated data for 300 hypothetical participants over 15 observed time points using the measurement model:

$$\begin{bmatrix} y_{i1}(t) \\ y_{i2}(t) \\ y_{i3}(t) \\ y_{i4}(t) \\ y_{i5}(t) \\ y_{i6}(t) \end{bmatrix} = \begin{bmatrix} =1 & 0 & 1 & 0 \\ \lambda_{21} & 0 & 1 & 0 \\ \lambda_{31} & 0 & 1 & 0 \\ 0 & =1 & 1 & 0 \\ 0 & \lambda_{52} & 1 & 0 \\ 0 & \lambda_{62} & 1 & 0 \end{bmatrix} \begin{bmatrix} f_{i1}(t) \\ f_{i2}(t) \\ setpoint_i(t) \\ \Delta setpoint_i(t) \end{bmatrix}, \tag{7.18}$$

in which the first factor loading of each latent factor was fixed at 1.0 for identification purposes; $setpoint_i(t)$ denotes the local level of the common trend at time t, and $\Delta setpoint_i(t)$ is the corresponding local slope that determines the magnitude and direction of the changes in common trend from one time point to the next. Here, we generated sigmoid-shaped trajectories for $setpoint_i(t)$ and VAR(1) dynamics for the other latent factors as

$$\begin{bmatrix} f_{i1}(t) \\ f_{i2}(t) \\ setpoint_i(t) \\ \Delta setpoint_i(t) \end{bmatrix} = \begin{bmatrix} \phi_{11} & \phi_{12} & 0 & 0 \\ \phi_{21} & \phi_{22} & 0 & 0 \\ 0 & 0 & \beta & 1 \\ 0 & 0 & 0 & 1 \end{bmatrix} \begin{bmatrix} f_{i1}(t-1) \\ f_{i2}(t-1) \\ setpoint_i(t-1) \\ \Delta setpoint_i(t-1) \end{bmatrix} + \begin{bmatrix} \zeta_{i1}(t) \\ \zeta_{i2}(t) \\ 0 \\ \zeta_{i\Delta setpoint}(t) \end{bmatrix}, \tag{7.19}$$

with initial conditions:

$$\begin{bmatrix} f_{i1}(1) & f_{i2}(1) & setpoint_i(1) & \Delta setpoint_i(1) \end{bmatrix}^T \sim N\left(\mathbf{0}, \text{diag}\begin{bmatrix} 1 & 1 & 0.5 & 0.2 \end{bmatrix}\right). \tag{7.20}$$

The equation for *setpoint* is specified to be dependent on its previous level at time $t - 1$ with β set to 0.8, and a constant slope, $\Delta setpoint_i(t)$, that remains invariant over time. This model, known as the dual change score model, was used by McArdle and Hamagami (2001) to represent sigmoid-shaped changes in human fluid and crystallized intelligence from early childhood to adulthood.

The true parameters were set to values that ensured stable dynamics: $\phi_{11} = 0.8$, $\phi_{12} = 0$, $\phi_{21} = -0.2$, and $\phi_{22} = 1$. Other true parameter values and details on simulation settings can be found in the demonstrative code included with this chapter. Using the model shown in Equations (7.18)–(7.20), we obtained the trajectories for $fs_{i1}(t)$, $setpoint_i(t)$, $\Delta setpoint_i(t)$, and $y_{i1}(t)$ as shown in Figure 7.2A–D, respectively.

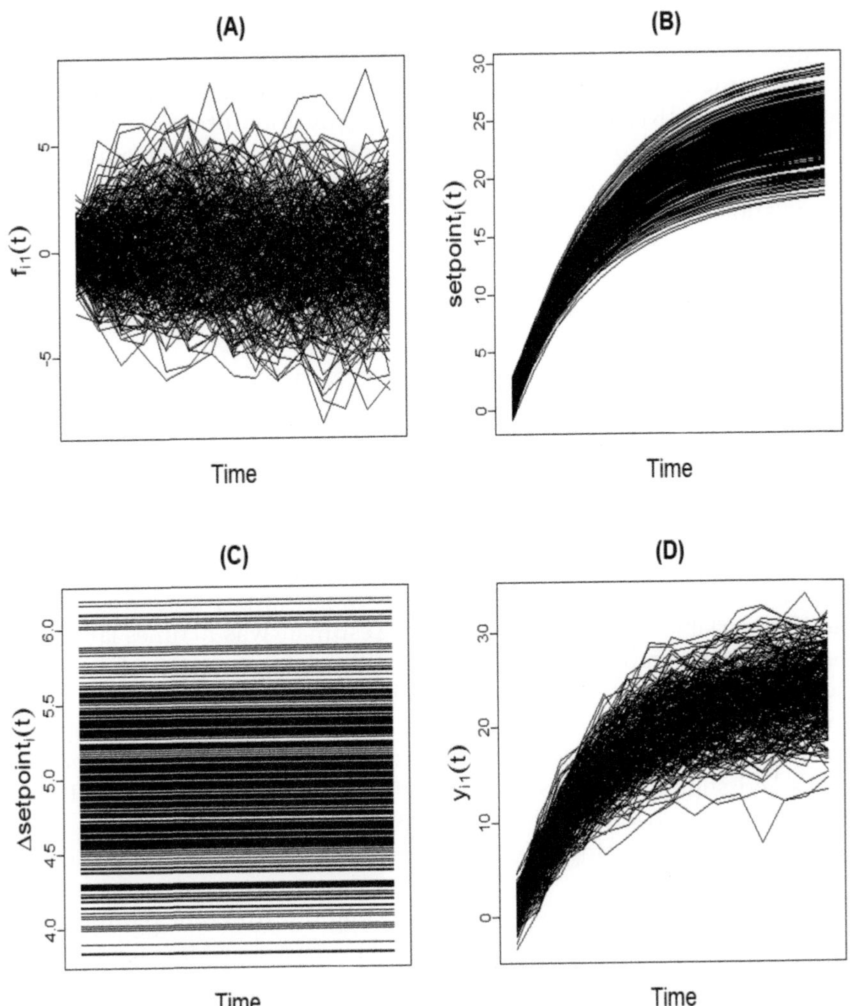

FIGURE 7.2
Trajectories of: (A) $f_{i1}(t)$, (B) $setpoint_i(t)$, (C) $\Delta setpoint_i(t)$, and (D) $y_{i1}(t)$ generated using Equations (7.18)–(7.20).

Based on a priori expectations for a prominent learning trend, one may use, for instance, the IRW as a model of choice for approximating the dynamics of the TV set-point. If the learning trend is expected to be smooth (as was the case in this illustration), the SIRW may be preferred. Here, we considered estimates from five possible models for representing the time evolution of the set-point: (1) the RW model; (2) the IRW with initial conditions fixed at the true means and variance values; (3) the IRW with freely estimated initial condition means and variances; (4) the SIRW with initial conditions fixed at the true means and variance values; and (5) a model with time-invariant set-point, obtained as a special case of the RW model in which the variance of $\zeta_{i\Delta setpoint}(t)$ was set to be zero.

We assumed a diagonal covariance structure for all the ICs. However, the ICs for the IRW model involve the specification of average and individual differences in initial level and slope. Information from multiple individuals across at least three time points is required to identify the means and variances of the initial level and slope, as well as their covariance. Thus, rather than freely estimating the IC-related parameters in the IRW model, we recommend using alternative approaches to either fix or estimate the ICs in the IRW model (Du Toit & Browne, 2007; Harvey, 2001; Ji & Chow, 2018).

To proceed, we considered a heuristic approach in this example. We specified the means and variances of the ICs, $f_{i1}(1)$, $f_{i2}(1)$, and initial level of the setpoint ($\vartheta_i(1)$ in Equation (7.15)), as freely estimated parameters. As distinct from this specification, we set the mean of $\Delta_i(1)$, the initial local slope for setpoint, to the empirical mean of the differences in the six manifest variables between time 1 and time 2 across all individuals and manifest variables, and the variance to be the corresponding variance (across individuals) in empirical mean differences between time 1 and time 2 across manifest variables. Note that this variance estimate was 10 times larger than the true initial condition variance used in the data generation. Our experience suggests that it is preferable to allow for a relatively diffuse (rather than precise) specification of the initial conditions. The diffuse specification comes at the price of less accurate estimates of the initial latent variable values and slightly less precise parameter estimates (as illustrated below). However, doing so still yields much improved estimation results compared to the alternative of using overly restrictive but inaccurate IC specifications (Ji & Chow, 2018).

The five models for the set-point have slightly different parameters in the functions for the TVP. Thus, we focus on comparing biases in the latent variable estimates of $f_{i1}(t)$, $f_{i2}(t)$, and $setpoint_i(t)$; and parameter estimates involving functions of the two latent factors, $f_{i1}(t)$, and $f_{i2}(t)$ (see Table 7.2). The estimates from all models were reasonably satisfactory, except for those associated with the time-invariant model. In particular, incorrectly specifying TV set-points as time-invariant led to considerable biases in estimates of the auto- and cross-regression parameters. Notable biases were also observed in the smoothed estimates of the set-point from the time-invariant model.

The SIRW model, on average, showed the best performance in estimating values of the latent variables and parameters (see Table 7.3). This is not surprising, because the SIRW model includes as a special case the true data generation model shown in Equations (7.12)–(7.14). Biases in parameter and latent variable estimates were, on average, similar in the two variations of the IRW model, regardless of whether the IC parameters were fixed to their true values or freely estimated. Inspection of the estimates of the latent variable

TABLE 7.2

Parameter Estimates for Illustrative Example 1

	True	RW	IRW (Fixed IC)	IRW (Free IC)	SIRW	Invariant
ϕ_{11}	0.80	0.81	0.80	0.79	0.80	0.66
ϕ_{21}	−0.20	−0.23	−0.24	−0.25	−0.21	−0.09
ϕ_{12}	0.00	0.02	0.01	−0.00	0.01	0.15
ϕ_{22}	0.70	0.69	0.69	0.68	0.69	0.89
λ_{21}	1.50	1.53	1.54	1.54	1.49	1.02
λ_{31}	1.00	1.01	1.00	1.00	1.00	1.01
λ_{52}	0.80	0.81	0.79	0.80	0.80	1.00
λ_{62}	1.00	1.00	1.01	1.01	1.01	1.00
ψ_{11}	2.00	1.48	1.74	1.76	1.97	3.21
ψ_{12}	0.50	0.02	0.32	0.33	0.51	0.75
ψ_{22}	1.50	0.99	1.33	1.34	1.51	1.60
$Var(\epsilon_1)$	0.50	0.48	0.48	0.49	0.49	1.03
$Var(\epsilon_2)$	0.50	0.58	0.47	0.48	0.51	2.16
$Var(\epsilon_3)$	0.50	0.54	0.53	0.53	0.53	0.00
$Var(\epsilon_4)$	0.50	0.53	0.53	0.53	0.53	0.60
$Var(\epsilon_5)$	0.50	0.49	0.48	0.48	0.48	0.49
$Var(\epsilon_6)$	0.50	0.49	0.48	0.48	0.48	0.55

scores for one randomly selected participant in Figures 7.3A–C indicated that more notable biases in estimates of the set-point occurred primarily during the initial stretches of the data. This bias reduced substantially after approximately five measurement occasions. The RW and IRW models both yielded estimates of the latent variable scores that closely approximated the

TABLE 7.3

Biases in Latent Variable Biases for Illustrative Example 1

	Average Bias $f_{i1}(t)$	Average Bias $f_{i2}(t)$	Average Bias $setpoint_i(t)$
RW	−0.04	−0.04	0.04
IRW fixed IC	−0.01	−0.02	0.01
IRW free IC	−0.03	−0.04	0.03
Smoothed IRW	0.00	0.01	−0.00
Invariant	−7.94	−7.99	7.98

Note: Average biases across all parameters from the five models: −0.08 (RW model), −0.04 (IRW model with fixed and correctly specified IC), −0.04 (IRW model with freely estimated IC), 0.00 (SIRW model), and 0.20 (time-invariant model).

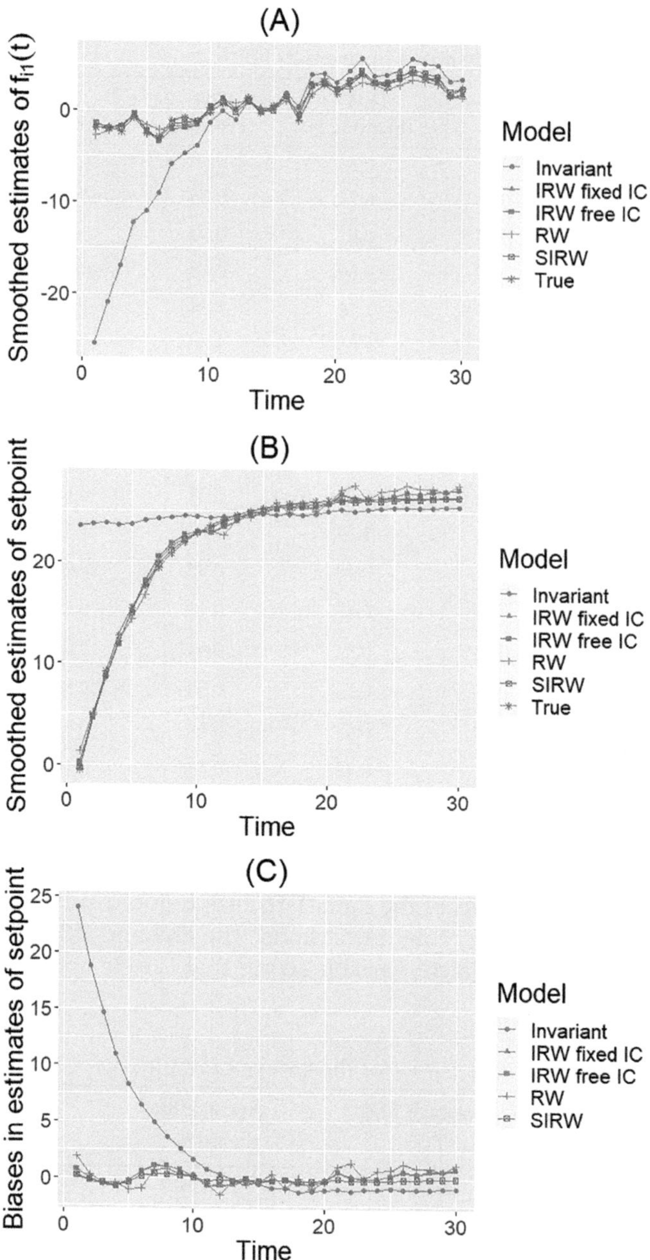

FIGURE 7.3
(A): Plots of true and estimated trajectories of fi1(t). (B) and (C): Plots of true and estimated trajectories of $setpoint_i(t)$ and the corresponding biases. RW = random walk model; IRW = integrated random walk model; True = true trajectory; Invariant = estimates from the model with time-invariant set-point.

true variations in the set-point, except that the RW tended to overestimate extreme curvatures in the data.

7.6.2 DFM(p,q,0,1,0) with Time-Varying Set-Point and Cross-Regression Parameters

We used a set of previously published empirical data (Chow & Zhang, 2013) to illustrate the consequences of ignoring the systematic trends in the data. The illustrative data set was a subset of the data from the Affective Dynamics and Individual Differences study. Participants between 18 and 86 years old enrolled in a laboratory study of emotion regulation, followed by an experience sampling study during which the participants rated their momentary feelings five times daily close to, or slightly over a month. Only the experience sampling data were used in the present analysis. After removing the data of participants with excessive missingness (> 65% missingness) and data that lacked sufficient response variability, data of 217 participants were included in the final sample.

The two key latent variables of interest were individuals' positive affect (PA) and negative affect (NA), measured using items from the Positive Affect and Negative Affect Schedule (Watson, Clark, & Tellegen, 1988) and other items posited in the circumplex model of affect (Larsen & Diener, 1992; Russell, 1980) on a scale of 1 (never) to 4 (very often). As in Chow and Zhang (2013), we created three item parcels as indicators of each of the two latent factors (PA and NA), respectively, via item parceling (Cattell & Barton, 1974; Kishton & Widaman, 1994). The items included in the three parcels included: (1) for PA parcel 1, elated, affectionate, lively, attentive, active, satisfied, and calm; (2) for PA parcel 2, excited, love, enthusiastic, alert, interested, pleased, and happy; (3) for PA parcel 3, aroused, inspired, proud, determined, strong, and relaxed; (4) for NA parcel 1, angry, sad, distressed, jittery, guilty, and afraid; (5) for NA parcel 2, upset, hostile, irritable, tense, and ashamed, and (6) for NA parcel 3, depressed, agitated, nervous, anxious, and scared. All items were assessed five times daily at partially randomized intervals that included both daytime assessments as well at least one assessment in the evening. Consistent with the data pre-processing step taken by Chow and Zhang (2013), we aggregated the composite scores over every 12-hour block to yield two measurements per day up to 37 days (a few of the participants continued to provide responses beyond the requested one-month study period). The total number of time points for each participant ranged from 26 to 74 time points, with an average missing data proportion of 0.18.

Chow et al. (2011) used a theoretical model of affect, namely, the Dynamic Model of Activation proposed by Zautra and colleagues (Reich, Zautra, & Davis, 2003; Zautra et al., 2000; Zautra, Potter, & Reich, 1997), to construct and test a dynamic factor analysis model with TV cross-regression relations at the factor level using a different data set. According to the Dynamic Model of Activation (Reich et al., 2003; Zautra et al., 1997, 2000), PA and NA are independent under low-stress (activation) conditions. However, they tend to collapse into a unidimensional, bipolar structure under high activation. Chow et al. (2011) sought to clarify and further disentangle the directionality of the PA-NA linkage by considering how the lead-lag relationship between PA and NA changed over time in the context of a DFM$(p,q,0,1,0)$ model (Browne & Nesselroade, 2005; McArdle, 1982) with vector autoregressive relation of order 1 at the factor level. As distinct from the standard DFM$(p,q,0,1,0)$ model, they allowed the cross-regression parameters between the two latent factors, PA and NA, to vary over time following an AR of differences model:

$$\begin{aligned}
PA_i(t) &= \phi_{11}PA_i(t-1)+\phi_{i,12}(t)NA_i(t-1)+\zeta_{1i}(t)\\
NA_i(t) &= \phi_{22}NA_i(t-1)+\phi_{i,21}(t)PA_i(t-1)+\zeta_{2i}(t)\\
\phi_{i,12}(t)-\phi_{120} &= \alpha_{12}\left[\phi_{i,12}(t-1)-\phi_{120}\right]+\zeta_{i,NA\rightarrow PA}(t)\\
\phi_{i,21}(t)-\phi_{210} &= \alpha_{21}\left[\phi_{i,21}(t-1)-\phi_{210}\right]+\zeta_{i,PA\rightarrow NA}(t).
\end{aligned} \tag{7.21}$$

This alternative DFM model with TV cross-regression parameters posits that individuals' cross-regression parameters fluctuate around the baseline values, ϕ_{120} and ϕ_{210}, following an AR(1) model. However, Chow et al. did not consider a simpler possibility: that the TV relations between PA and NA were artifacts of trends in individuals' PA and NA over time. That is, even though we did remove individual- and variable-specific means and linear trends from the data prior to further dynamic modeling, other nonlinear trends might still be present in the data and could affect modeling results. Thus, we compared the fit of the model considered by Chow et al. in Equation (7.21), denoted herein as the *TV-CR Model* with (1) a standard DFM$(6,2,0,1,0)$ model with time-invariant parameters, denoted herein as the *Time-Invariant DFM*; and (2) a DFM$(6,2,0,1,0)$ model with time-invariant cross-regression parameters and a common trend (i.e., TV set-point) characterized by the SIRW model, denoted herein as *DFM with Trend*, expressed as

$$\begin{aligned}
PA_i(t) &= setpoint_i(t-1)+\phi_{11}PA_i(t-1)+\phi_{i,12}(t)NA_i(t-1)+\zeta_{1i}(t)\\
NA_i(t) &= setpoint_i(t-1)+\phi_{22}NA_i(t-1)+\phi_{i,21}(t)PA_i(t-1)+\zeta_{2i}(t)
\end{aligned} \tag{7.21a}$$

The measurement model associated with the *TV-CR Model* in Equation (7.21) follows the form:

$$\begin{bmatrix} y_{i1}(t)\\ y_{i2}(t)\\ y_{i3}(t)\\ y_{i4}(t)\\ y_{i5}(t)\\ y_{i6}(t)\end{bmatrix}=\begin{bmatrix} =1 & 0 & 0 & 0\\ \lambda_{21} & 0 & 0 & 0\\ \lambda_{31} & 0 & 0 & 0\\ 0 & =1 & 0 & 0\\ 0 & \lambda_{52} & 0 & 0\\ 0 & \lambda_{62} & 0 & 0\end{bmatrix}\begin{bmatrix} PA_i(t)\\ NA_i(t)\\ \phi_{i,12}(t)\\ \phi_{i,21}(t)\end{bmatrix}; \tag{7.22}$$

whereas the measurement model for the *Time-Invariant DFM* and *DFM with Trend* is as shown in Equation (7.18).

Based on the AIC and BIC values from the three models, we preferred *TV-CR* Model (see Table 7.4). All parameters, except for ϕ_{210} and α_{21}, were significantly different from zero. We fixed these parameters at zero and re-estimated the model, denoted as "Final TV cross-reg" in Table 7.4. This "final" model was characterized by the best AIC and BIC values among all the models considered for this empirical illustration. All subsequent discussions are based on this particular model.

TABLE 7.4

Parameter Estimates from the Empirical Example Using Data from the ADID Study

	DFM w/	Trend	Time-Invariant DFM		Final TV-CR Model	
	Est	SE	Est	SE	Est	SE
ϕ_{11}	0.42	0.02	0.39	0.01	0.36	0.01
ϕ_{21}	−0.03	0.01	−0.01	0.01	0.00	–
ϕ_{12}	−0.06	0.02	−0.05	0.01	−0.06[a]	0.02[a]
ϕ_{22}	0.42	0.02	0.48	0.01	0.49	0.01
λ_{21}	1.14	0.01	1.10	0.01	1.10	0.01
λ_{31}	0.77	0.01	0.82	0.01	0.82	0.01
λ_{52}	1.13	0.01	1.08	0.01	1.08	0.01
λ_{62}	1.14	0.01	1.10	0.01	1.10	0.01
ψ_{11}	0.05	0.00	0.07	0.00	0.06	0.00
ψ_{12}	−0.02	0.00	−0.01	0.00	−0.01	0.00
ψ_{22}	0.03	0.00	0.03	0.00	0.03	0.00
$\text{Var}(\epsilon_1)$	0.01	0.00	0.01	0.00	0.01	0.00
$\text{Var}(\epsilon_2)$	0.01	0.00	0.01	0.00	0.02	0.00
$\text{Var}(\epsilon_3)$	0.03	0.00	0.03	0.00	0.03	0.00
$\text{Var}(\epsilon_4)$	0.01	0.00	0.01	0.00	0.01	0.00
$\text{Var}(\epsilon_5)$	0.02	0.00	0.02	0.00	0.02	0.00
$\text{Var}(\epsilon_6)$	0.02	0.00	0.02	0.00	0.02	0.00
	AIC	BIC	AIC	BIC	AIC	BIC
	31255.32	31098.44	31283.16	31126.28	31662.18	31520.24

Note: *DFM w/Trend* = DFM(6,2,0,1,0) model with SIRW representation of a common trend for PA and NA; *Time-Invariant DFM* = DFM(6,2,0,1,0) with time-invariant parameters; *Final TV-CR Model* = final DFM(6,2,0,1,0) model with TV CR parameters.

[a] In the final *TV-CR Model*, the point and SE estimates for ϕ_{12} shown here correspond to the estimates for the parameter ϕ_{120}. Additional parameter estimates from the final *TV-CR Model* (with SE estimates enclosed in parentheses) included: α_{12} = 0.87 (0.05); $\text{Var}(\zeta_{NA\rightarrow PA})$ = 0.03 (0.01); $\text{Var}(\zeta_{PA\rightarrow NA})$ = 0.07 (0.01). α_{21} was not significantly different from zero and was omitted from the final model.

Results from the final TV cross-regression model suggest that there was a systematic and statistically significant cross-regression influence in the direction of NA to PA, but not in the direction of PA to NA, as evidenced by the statistically significant estimated values of ϕ_{120}. Some cross-regression influence was present in the direction of NA to PA, but this influence tended to ebb and flow in a noise-like manner, as driven only by process noise and did not show systematic continuity over time. Estimated values of these now TV cross-regression parameters are shown in Figure 7.4A–B.

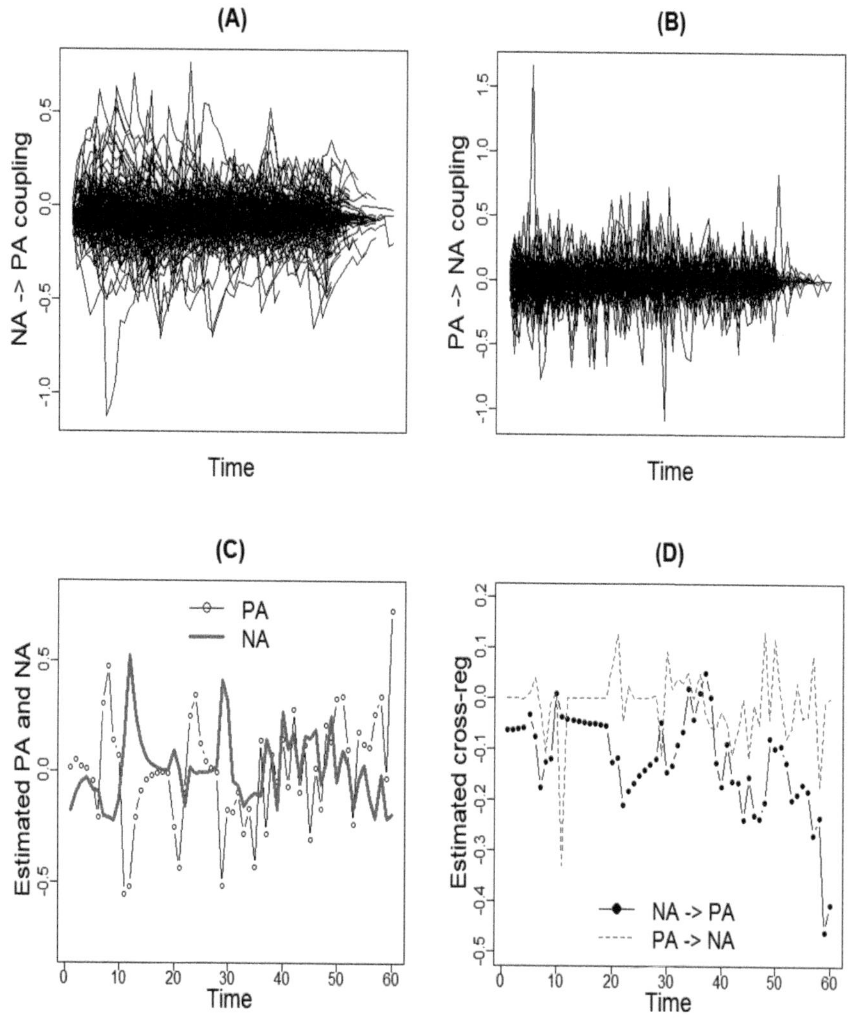

FIGURE 7.4
(A) and (B): Plot of the estimated trajectories of *setpoint*$_i$*(t)* for PA and NA, respectively, based on the SIRW model. (C) and (D): Plot of the estimated trajectories of $\phi_{i,12}(t)$ and $\phi_{i,21}(t)$, respectively, for the NA → PA and PA → NA cross-regressions for one randomly selected participant.

To further illustrate the potential implications of such TV cross-regression relations, we plotted the estimated values of one randomly selected participant's estimated PA and NA factor scores, as well as the participant's estimated cross-regression parameters over time (see Figure 7.4C–D). Based on these plots, we can see, for instance, in Figure 7.4C that between the 10th and the 20th time point, PA and NA were observed to show inversely related fluctuations in levels—that is, as NA showed sudden surges to high (above 0) values, PA was observed to be low (below 0) in values. The estimated over-time dynamics of the two cross-regression parameters indicate that the cross-regression influence from PA to NA remain relatively constant and close to zero during this period, whereas the cross-regression influence from NA to PA became progressively more negative during this window of time. Thus, when this individual's NA at the previous time point was high (assuming a large positive value), PA at the current time point tended to be lower (assuming a larger negative value) than expected if the individual's previous NA was at the average level (i.e., at 0).

In contrast, during the last 20 time points, estimates of PA and NA showed notable upward and downward trends of increases and decreases, respectively, and some fluctuations around these trends. These changes were driven by changes in both the PA → NA as well as NA → PA cross-regression parameters. In particular, divergent changes in the levels of PA and NA were observed to occur when the NA → PA cross-regression influence assumed increasingly larger negative values. That is, the high levels of PA during this time window were closely related to, or possibly driven by the sustained low levels of low NA on the preceding occasions. In addition to these over-time variations in the NA → PA cross-regression influence, there were times (e.g., at $t = 50$) when the PA → NA cross-regression parameter became positive in value. These changes resulted in PA and NA showing synchronous changes in levels (e.g., simultaneous increases in levels) that were distinct from their typically divergent trends.

To conclude, results from model fitting confirmed the presence of nonstationarity or instability in the dynamics of individuals' PA and NA over time. The better fit of the *TV-CR Model* relative to the *DFM with Trend* indicates that there are fundamental within-person variations in the linkages between PA and NA over time that are not due to the shared trend between PA and NA. One important "take-home message" from these results is that one may arrive at fundamentally different conclusions concerning the nature of individuals' affect structure if parameters that characterize how the processes of interest change over time also vary over time. For instance, drawing on results from the model with time-invariant parameters, one might conclude that the links between PA and NA were driven solely by the influence of NA to PA. This is only partially true. We found that there was some cross-regression influence from PA to NA, though this influence tended to be transient and noise-like in nature.

7.7 Closing Remarks

As we have illustrated, dynamic models with varying coefficients provide a natural platform for testing certain psychological processes of interest. We presented one possible approach for examining time-dependencies in modeling parameters. This is by no means the only approach. For instance, parameters in a model may show dependencies not on time but on other covariates such as space and contextual factors. One alternative approach for probing such dependencies is to use estimation procedures from the generalized additive modeling framework (Bringmann et al., 2017; Hastie, 2015; Tan et al., 2012; Wood, 2017) to examine if and how modeling parameters show dependencies on measured covariates within a general or generalized linear modeling framework. The generalized additive modeling framework does not readily allow for the inclusion of latent variables. However, it has the advantages of using spline functions to allow for greater flexibility in representing the nature of the varying coefficients, and penalized estimation routines to shrink unimportant coefficients and simplify model structures. The details of such an approach are beyond the scope of this chapter.

Even though many phenomena in the social and behavioral sciences do in fact show nonstationarity in the form of varying coefficients, they are rarely put to test using formal model-fitting and model comparison approaches. More often, carefully construed experiments are designed to explicitly impose expected changes in patterns of intra-individual variability within and across conditions. The lack of readily accessible tools for evaluating more complex dynamic models is one reason for the scarcity of modeling work along this line. We hope to have provided the reader with a workable set of examples showcasing the potential utility and promise of models with varying coefficients.

References

Anderson, B.D.O. & Moore, J.B. (1979). *Optimal filtering*. Englewood Cliffs, NJ: Prentice Hall.

Ansley, C.F. & Kohn, R. (1985). Estimation, filtering and smoothing in state space models with incompletely specified initial conditions. *The Annals of Statistics, 13*(4), 1286–1316.

Bar-Shalom, Y., Li, X.R., & Kirubarajan, T. (2001). *Estimation with applications to tracking and navigation: Theory algorithms and software*. New York, NY: Wiley.

Bringmann, L.F., Hamaker, E.L., Vigo, D.E., Aubert, A., Borsboom, D., & Tuerlinckx, F. (2017). Changing dynamics: Time-varying autoregressive models using generalized additive modeling. *Psychological Methods, 22*(3), 409–425.

Browne, M.W. (1993). Structured latent curve models. In C.M. Cuadras & C.R. Rao (Eds.), *Multivariate analysis: Future directions 2* (pp. 171–198). Amsterdam: North-Holland.

Browne, M.W. & du Toit, H.C. (1991). Models for learning data. In L.M. Collins & J.L. Horn (Eds.), *Best methods for the analysis of change: Recent advances, unanswered questions, future directions* (pp. 47–68). Washington, DC: American Psychological Association.

Browne, M.W. & Nesselroade, J.R. (2005). Representing psychological processes with dynamic factor models: Some promising uses and extensions of autoregressive moving average time series models. In A. Maydeu-Olivares & J.J. McArdle (Eds.), *Contemporary psychometrics: A Festschrift for Roderick P. McDonald* (pp. 415–452). Mahwah, NJ: Erlbaum.

Caines, P.E. & Rissanen, J. (1974). Maximum likelihood estimation of parameters. *IEEE Transactions on Information Theory, 20*(1), 102–104.

Cattell, R. & Barton, K. (1974). Changes in psychological state measures and time of day. *Psychological Reports, 35*(1), 219–222.

Cattell, R.B., Cattell, A.K.S., & Rhymer, R.M. (1947). P-technique demonstrated in determining psychophysical source traits in a normal individual. *Psychometrika, 12*(4), 267–288.

Chen, M., Chow, S.-M., Hammal, Z., Messinger, D.S., & Cohn, J.F. (2020). A person- and time-varying vector autoregressive model to capture interactive infant-mother head movement dynamics. *Multivariate Behavioral Research, 56*(5), 739–767. doi: 10.1080/00273171.2020.1762065.

Chow, S.-M., Ferrer, E., & Nesselroade, J.R. (2007). An unscented Kalman filter approach to the estimation of nonlinear dynamical systems models. *Multivariate Behavioral Research, 42*(2), 283–321.

Chow, S.-M., Ho, M.-H.R., Hamaker, E.J., & Dolan, C.V. (2010). Equivalences and differences between structural equation and state-space modeling frameworks. *Structural Equation Modeling, 17*(2), 303–332.

Chow, S.-M., Lu, O., & Cohn, J.F., & Messinger, D.S. (2016). Representing self- organization and non-stationarities in dyadic interaction processes using dynamic systems modeling techniques. In A. Von Davier, P. C. Kyllonen, & M. Zhu (Eds.), *Innovative assessment of collaboration*. New York, NY: Springer.

Chow, S.-M. & Zhang, G. (2013). Nonlinear regime-switching state-space (RSSS) models. *Psychometrika: Application Reviews and Case Studies, 78*(4), 740–768.

Chow, S.-M., Zu, J., Shifren, K., & Zhang, G. (2011). Dynamic factor analysis models with time-varying parameters. *Multivariate Behavioral Research, 46*(2), 303–339.

Craigmile, P.F., Peruggia, M., & Van Zandt, T. (2009). Detrending response time series. In S.-M. Chow, E. Ferrer, & F. Hsieh (Eds.), *Statistical methods for modeling human dynamics: An interdisciplinary dialogue* (pp. 213–240). New York, NY: Taylor & Francis.

De Jong, P. (1991). The diffuse Kalman filter. *The Annals of Statistics, 19*(5), 1073–1083. doi: 10.1214/aos/1176348139.

Doucet, A., de Freitas, N., & Gordon, N. (Eds.). (2001). *Sequential Monte Carlo methods in practice*. New York, NY: Springer. doi: 10.1007/978-1-4757-3437-9.

Drozdick, L.W., Wahlstrom, D., Zhu, J., & Weiss, L.G. (2012). The Wechsler Adult Intelligence Scale—Fourth Edition and the Wechsler Memory Scale—Fourth Edition. InD.P. Flanagan & P.L. Harrison (Eds.), *Contemporary intellectual assessment: Theories, tests, and issues* (pp. 197–223). New York, NY: The Guilford Press.

Durbin, J. & Koopman, S.J. (2001). *Time series analysis by state space methods*. New York, NY: Oxford University Press.

Du Toit, S.H.C. & Browne, M.W. (2007). Structural equation modeling of multivariate time series. *Multivariate Behavioral Research, 42*(1), 67–101.

Fan, J. & Zhang, W. (2000). Simultaneous confidence bands and hypothesis testing in varying-coefficient models. *Scandinavian Journal of Statistics: Theory and Applications*, 27(4), 715–731. doi: 10.1111/1467-9469.00218.

Fan, J. & Zhang, W. (2008). Statistical methods with varying coefficient models. *Statistics and Its Interface, 1*(1), 179–195.

Gelb, A. (1974). *Applied optimal estimation*. Cambridge, MA: MIT Press.

Haken, H., Kelso, J.A.S., & Bunz, H. (1985). A theoretical model of phase transitions in human hand movements. *Biological Cybernetics, 51*(5), 347–356. doi: 10.1007/BF00336922.

Haken, H. (2006). Pattern recognition and synchronization in pulse-coupled neural networks. *Nonlinear Dynamics, 44*(1), 269–276.

Harvey, A.C. (2001). *Forecasting, structural time series models and the Kalman filter*. Cambridge: Cambridge University Press.

Harvey, A.C. & Phillips, G.D.A. (1982). The estimation of regression models with time-varying parameters. In M. Deistler, E. Fürst, & G. Schwödiauer (Eds.), *Games, economic dynamics, and time series analysis: A symposium in memoriam Oskar Morgenstern organized at the Institute for Advanced Studies, Vienna* (pp. 306–321). Heidelberg: Physica-Verlag HD. doi: 10.1007/978-3-662-41533-7

Hastie, T. (2015). GAM: Generalized additive models [Computer software manual]. Retrieved from http://CRAN.R-project.org/package=gam (R package version 1.12).

Hastie, T. & Tibshirani, R. (1993). Varying-coefficient models. *Journal of the Royal Statistical Society Series B (Methodological), 55*(4), 757–796.

Jakeman, A.J. & Young, P.C. (1979). Recursive filtering and the inversion of ill-posed causal problems, CRES Report No. AS/R28/1979 (Tech. Rep.). Centre for Resource and Environmental Studies, Australian National University.

Jakeman, A.J. & Young, P.C. (1984). Recursive filtering and the inversion of ill-posed causal problems. *Utilitas Mathematica, 35*, 351–376.

Ji, L. & Chow, S.-M. (2018). Methodological issues and extensions to the latent difference score framework. In E. Ferrer, S. M. Boker, & K. J. Grimm (Eds.), *Longitudinal multivariate psychology* (pp. 9–37). New York, NY: Taylor & Francis. doi: 10.4324/9781315160542-2.

Kishton, J.M. & Widaman, K.F. (1994). Unidimensional versus domain representative parceling of questionnaire items: An empirical example. *Educational and Psychological Measurement, 54*(3), 757–765. doi: 10.1177/0013164494054003022.

Kitagawa, G. (1998). A self-organizing state-space model. *Journal of the American Statistical Association, 93*(443), 1203–1215.

Koopman, S.J. (1997). Exact initial Kalman filtering and smoothing for nonstationary time series models. *Journal of the American Statistical Association, 92*(440), 1630–1638.

Larsen, R.J. & Diener, E. (1992). Promises and problems with the circumplex model of emotion. *Review of Personality and Social Psychology, 13*, 25–59.

Lütkepohl, H. (2005). *Introduction to multiple time series analysis* (2nd ed.). New York, NY: Springer-Verlag.

MacCallum, R.C. & Ashby, G.F. (1986). Relationships between linear systems theory and covariance structure modeling. *Journal of Mathematical Psychology, 30*(1), 1–27. doi: 10.1016/0022-2496(86)90039-8.

McArdle, J.J. (1982). Structural equation modeling of an individual system: Preliminary results from "A case study in episodic alcoholism" (Unpublished manuscript, Department of Psychology, University of Denver).

McArdle, J.J. & Epstein, D.B. (1987). Latent growth curves within developmental structural equation models. *Child Development, 58*(1), 110–133.

McArdle, J.J. & Hamagami, F. (2001). Latent difference score structural models for linear dynamic analysis with incomplete longitudinal data. In L. Collins & A. Sayer (Eds.), *New methods for the analysis of change* (pp. 139–175). Washington, DC: American Psychological Association.

Meredith, W. (1993). Measurement invariance, factor analysis and factor invariance. *Psychometrika, 58*(4), 525–543.

Meredith, W. & Tisak, J. (1990). Latent curve analysis. *Psychometrika, 55*(1), 107–122.

Molenaar, P.C.M. (1987). Dynamic assessment and adaptive optimization of the psychotherapeutic process. *Behavioral Assessment, 9*(4), 389–416.

Molenaar, P.C.M. (1994). Dynamic latent variable models in developmental psychology. In A. von Eye & C. Clogg (Eds.), *Latent variables analysis: Applications for developmental research* (pp. 155–180). Thousand Oaks, CA: Sage Publications.

Molenaar, P.C.M., de Gooijer, J.G., & Schmitz, B. (1992). Dynamic factor analysis of nonstationary multivariate time series. *Psychometrika, 57*(3), 333–349.

Molenaar, P.C.M. & Nesselroade, J.R. (2009). The recoverability of P-technique factor analysis. *Multivariate Behavioral Research, 44*(1), 130–141. doi: 10.1080/00273170802620204.

Molenaar, P.C.M. & Newell, K.M. (2003). Direct fit of a theoretical model of phase transition in oscillatory finger motions. *British Journal of Mathematical and Statistical Psychology, 56*(2), 199–214. doi: 10.1348/000711003770480002.

Nesselroade, J.R. & Ford, D.H. (1985). P-technique comes of age multi-variate, replicated, single-subject designs for research on older adults. *Research on Aging, 7*(1), 46–80. doi: 10.1177/0164027585007001003.

Otter, P. (1986). Dynamic structure systems under indirect observation: Indentifiability and estimation aspects from a system theoretic perspective. *Psychometrika, 51*(3), 415–428.

Ou, L. (2018). Estimation of mixed effects continuous-time models [PhD Thesis]. Pennsylvania State University.

Ou, L., Hunter, M.D., & Chow, S.-M. (2019). What's for dynr: A package for linear and nonlinear dynamic modeling in R. *The R Journal.* https://journal.r-project.org/archive/2019/RJ-2019-012/index.html.

Oud, J.H.L. & Jansen, R.A.R.G. (2000). Continuous time state space modeling of panel data by means of SEM. *Psychometrika, 65*(2), 199–215.

Pagan, A. (1980). Some identification and estimation results for regression models with stochastically varying coefficients. *Journal of Econometrics, 13*(3), 341–363. doi: 10.1016/0304-4076(80)90084-6.

Reich, J.W., Zautra, A.J., & Davis, M. (2003). Dimensions of affect relationships: Models and their integrative implications. *Review of General Psychology, 7*(1), 66–83. doi: 10.1037/1089-2680.7.1.66.

Rocha, L.M. (1998). Selected self-organization and the semiotics of evolutionary systems. In G. van de Vijver, S. N. Salthe, & M. Delpos (Eds.), *Evolutionary systems: Biological and epistemological perspectives on selection and self-organization* (pp. 341–358). Dordrecht: Springer. doi: 10.1007/978-94-017-1510-2_25.

Russell, J.A. (1980). A circumplex model of affect. *Journal of Personality and Social Psychology, 39*(1-sup-6), 1161–1178.

Schweppe, F. (1965). Evaluation of likelihood functions for Gaussian signals. *IEEE Transactions on Information Theory, 11*(1), 61–70.

Shumway, R.H. & Stoffer, D.S. (2000). *Time series analysis and its applications.* New York, NY: Springer-Verlag.

Tan, X., Shiyko, M.P., Li, R., Li, Y., & Dierker, L. (2012). A time-varying effect model for intensive longitudinal data. *Psychological Methods, 17*(1), 61–77.

Tarvainen, M.P., Georgiadis, S.D., Ranta-aho, P.O., & Karjalainen, P.A. (2006). Time-varying analysis of heart rate variability signals with Kalman smoother algorithm. *Physiological Measurement, 27*(3), 225–239. doi: 10.1088/0967-3334/27/3/00.

Watson, D., Clark, L.A., & Tellegen, A. (1988). Development and validation of brief measures of positive and negative affect: The PANAS scale. *Journal of Personality and Social Psychology, 54*(6), 1063–1070.

Wechsler, D. (1981). The psychometric tradition: developing the Wechsler adult intelligence scale. *Contemporary Educational Psychology.*

Wechsler, D. (2014). *Wechsler intelligence scale for children (5th ed.)* [Computer software manual]. Bloomington, MN: Pearson.

Weiss, A.A. (1985). The stability of the AR(1) process with an AR(1) coefficient. *Journal of Time Series Analysis, 6*(3), 181–186. doi: 10.1111/j.1467-9892.1985.tb00408.x

West, M. & Harrison, J. (1989). *Bayesian forecasting and dynamic models.* Berlin: Springer-Verlag.

Wood, S. (2017). *Generalized additive models: An introduction with R* (2nd ed.). New York, NY: Chapman and Hall/CRC.

Zautra, A.J., Potter, P.T., & Reich, J.W. (1997). The independence of affect is context-dependent: An integrative model of the relationship between positive and negative affect. In P. Lawton, K.W. Schaie, & P. Lawton (Eds.), *Annual review of gerontology and geriatrics: Vol 17. Focus on adult development* (pp. 75–103). New York, NY: Springer.

Zautra, A.J., Reich, J.W., Davis, M.C., Potter, P.T., & Nicolson, N.A. (2000). The role of stressful events in the relationship between positive and negative affects: Evidence from field and experimental studies. *Journal of Personality, 68*(5), 927–951.

Zhao, J.-Q. & Zhao, Y.-Y. (2020). Bootstrap bandwidth selection in time-varying coefficient models with jumps. *Communications in Statistics – Simulation and Computation, 51*(6), 1–14. doi: 10.1080/03610918.2019.1708930.

8

Control Theory Optimization of Dynamic Processes

The use of control theory to steer a system to stay as close as possible to the desired reference state is very prevalent in areas such as engineering and physics (Åström & Murray, 2008; Bellman, 1964; Goodwin, Seron, & de Don, 2005; Kwon & Han, 2005; Liu et al., 2010; Wang et al., 2014a). Perhaps the best-known everyday example of such an application is the cruise control system of a car. In this case, the car with the cruise control is the "system"; the controller (the cruise control) determines the external input, namely, the engine's throttle position, which governs the power delivered by the engine to minimize the car's deviations from the desired (reference or target) speed set by the driver.

Myriad dynamic processes in the social and behavioral sciences also share similar characteristics with control theory problems. That is, across a range of applications, researchers and practitioners often seek ways to optimize the dosages of external input (e.g., medication, prevention, intervention, and educational training) needed to minimize the discrepancies between a system's actual and target levels (Savi, van der Maas, & Maris, 2015). These benefits have been demonstrated in the work of Wang et al. (2014a), who designed and implemented real-world control theory applications to compute the optimal insulin input to control the glucose levels of diabetic patients. Molenaar (2010) demonstrated via numerical simulations the plausibility of using control theory computation to optimize psychotherapy dosages to maintain desired levels of treatment effectiveness. In a similar vein, Rivera, Pew, and Collins (2007) also utilized simulation studies to investigate the anticipated impact of intervention design choices in developing adaptive interventions.

Despite these exceptions, most applications of control theory principles in the social, behavioral, and health sciences have been limited thus far to theoretical conceptualizations. Optimal control has close connections to the concepts of homeostasis and feedback, which have played a long-standing role in psychology (Carver & Scheier, 1982). Unfortunately, feedback control has thus far served primarily as an inspiration for theoretical models of psychological processes, due in part to a lack of understanding of the underlying mechanics and potential utility of control theory optimization, and ways of implementing them in everyday applications. When a parametric dynamic systems model has been successfully fitted to a psychological process, and

DOI: 10.1201/9780429172649-8

the model of interest includes the effects of measured external input, then this model can be employed to optimally control the process to desired levels if the external input can be manipulated. This is an important pay-off of analysis of intra-individual variation, which requires some dedicated computations but otherwise comes at no extra cost in experimental efforts.

It is our aim in this chapter to provide a first glimpse of the rationale underlying optimal control theory and provide supplementary code to clarify the implementation of the associated computational procedures. We demonstrate, within the context of a vector autoregressive model, how a system's latent values can be more efficaciously driven toward a pre-defined target level using one specific kind of controller known as the linear quadratic controller (LQC). Examples of control input in other applications have included manipulable dosages of insulin (Wang et al., 2014b), calorie intake (Rivera, Pew, & Collins, 2007), and training durations in an educational app (Chow et al., 2022). An illustration is provided to demonstrate how optimized control input can affect the state values of a system.

8.1 Control Theory Optimization

From the state-space model discussed in Chapters 5 and 7, a researcher may include *exogenous* or *input* variables that can affect the values of the latent state processes. In the event where the values of these input variables can be "manipulated" or altered to change the values of the state processes, these variables are known as *control inputs*, denoted herein as a vector, $\mathbf{z}_i(t)$. That is, $\mathbf{z}_i(t)$ may consist of multiple input variables, each of which may exert a different extent of influence on the state values. As an example, if the state values represent individuals' blood glucose levels, possible input variables may include insulin, as well as individuals' physical activity levels.

The state-space model thus assumes the form of

$$\boldsymbol{\eta}_{SS,i}(t) = \boldsymbol{\Phi}_{SS}\boldsymbol{\eta}_{SS,i}(t-1) + \mathbf{G}_{SS}\mathbf{z}_i(t) + \boldsymbol{\zeta}_{SS,i}(t), \tag{8.1}$$

$$\mathbf{y}_i(t) = \boldsymbol{\Lambda}_{SS}\boldsymbol{\eta}_{SS,i}(t) + \boldsymbol{\varepsilon}_i(t), \tag{8.2}$$

$$\boldsymbol{\zeta}_{SS,i}(t) \sim N(\mathbf{0}, \Psi_{SS}); \quad \boldsymbol{\varepsilon}_i(t) \sim N(\mathbf{0}, \Theta)$$

where $\mathbf{G}_{SS}$ is a matrix of regression coefficients relating the control inputs to the latent states. As in Chapter 7, Equations (8.1) and (8.2) constitute a group-based state-space model in which parameters in elements such as $\boldsymbol{\Lambda}_{SS}$ and $\boldsymbol{\Phi}_{SS}$ are constrained to be invariant across people. In applications with an adequate number of time points from each individual (e.g., $T > 50$), model fitting can also be done at the individual level, as discussed in Chapters 2–5 of this book. In those cases, we omitted the person index, i, to ease the presentation.

A system is controllable when the $w \times (w+r)$ controllability matrix,

$$\mathbf{C} = \left[\begin{array}{ccccc} \mathbf{G}_{SS} & \mathbf{\Phi}_{SS}\mathbf{G}_{SS} & \mathbf{\Phi}_{SS}^{2}\mathbf{G}_{SS} & \ldots & \mathbf{\Phi}_{SS}^{w-1}\mathbf{G}_{SS} \end{array} \right] \tag{8.3}$$

has rank w. The supplementary code for this chapter provides an example of how to perform a controllability test in R —> in the statistical software, R.

Optimal values of $z_i(t)$ may be obtained by minimizing an appropriate cost function with respect to $z_i(t)$ and the target state values. In other words, the goal is to find the values of $z_i(t)$ that can minimize a cost function of choice. There are many possible approaches for determining the values of $z_i(t)$ for each t so future measured values of a dynamical system at time t+1 and beyond are close to some pre-defined performance criteria. The control algorithm we demonstrate in this chapter, LQC with State Feedback (abbreviated as LQC-State), is one special case of a class of control strategies called Receding Horizon or model predictive controllers. It is linear in terms of the underlying state-space model (Equations 8.1–8.2) linked to the controller, quadratic in the sense of the quadratic form of the cost function adopted in Equation (8.4), and receding horizon refers to the property that the input is sequentially computed utilizing a receding (moving) window of future state values. Alternative terminologies are sometimes used to describe different controllers in the context of Receding Horizon and model predictive control literature. Kwon and Han (2005) provided a didactic overview of some of these standard (albeit seemingly divergent) terminology and jargons, and formalized many related control techniques under the framework of receding horizon control techniques for state-space models. Our explanations and derivations closely follow the framework set forth by these authors.

Quadratic cost functions are among the most popular cost functions in control theory applications due to the existence of well-known analytic functions for computing optimal control input values. Implementation of the LQC involves computations of the values of the input variables in $z_i(t)$ at each t over a future control horizon between time t and $t+h$, where $h>1$ is called the control horizon. One popular form of the quadratic cost function is structured:

$$\begin{aligned} J_i(t)\left(\boldsymbol{\eta}_{SS,i}(t+\tau)^{\{\tau=0,\ldots,h\}},\boldsymbol{\eta}_i^r,\mathbf{z}_i(t)\right) = \sum_{\tau=0}^{h-1}\Big[(\boldsymbol{\eta}_{SS,i}(t+\tau)-\boldsymbol{\eta}_i^r)'\mathbf{Q}(\boldsymbol{\eta}_{SS,i}(t+\tau)-\boldsymbol{\eta}_i^r) \\ +\mathbf{z}_i(t+\tau)'\mathbf{R}\mathbf{z}_i(t+\tau)\Big]+c_i(t+h), \end{aligned} \tag{8.4}$$

where $\boldsymbol{\eta}_i^r$ is a vector comprising the desired (possibly person-specific) reference levels of the latent processes, and $c_i(t+h) \overset{\Delta}{=} (\boldsymbol{\eta}_{i,t+h}-\boldsymbol{\eta}_i^r)'\mathbf{Q}_h(\boldsymbol{\eta}_{i,t+h}-\boldsymbol{\eta}_i^r)$. Thus, this cost function considers two types of "costs": costs in terms of deviations of individuals' state values from the target levels, and costs involved in administering the input.

The matrices $\mathbf{Q}$, $\mathbf{Q}_h$, and $\mathbf{R}$ are positive-definite design matrices chosen *a priori* to determine, respectively, how heavily three types of costs are penalized: (1) deviations of the latent processes from their desired levels within the control horizon (for τ between 0 and $h-1$); (2) deviations of the latent processes from their desired levels at the end point of the control horizon (i.e., for $\tau = h$); and (3) the costs associated with administration of $z_i(t)$. The actual magnitudes of $\mathbf{Q}$ (as well as $\mathbf{Q}_h$) and $\mathbf{R}$ are often determined in a relative sense in comparison to each other. For instance, suppose two manifest indicators are used to identify two latent processes, with their respective loadings fixed at unity, yielding a 2x2 identity matrix, $\Lambda_{SS} = \mathbf{I}_2$. One way to set $\mathbf{Q}$ and $\mathbf{Q}_h$ might be to set

$$\mathbf{Q} = \mathbf{Q}_h = \Lambda_{SS}\Lambda_{SS} = \begin{bmatrix} 1 & 0 \\ 0 & 1 \end{bmatrix}\begin{bmatrix} 1 & 0 \\ 0 & 1 \end{bmatrix} = \begin{bmatrix} 1 & 0 \\ 0 & 1 \end{bmatrix}.$$

This specification implies that deviations in both of the latent variables are penalized as heavily as each other. Then, the magnitudes $\mathbf{R}$ could be set relative to the unity weights in $\mathbf{Q}$ and $\mathbf{Q}_h$. For instance, suppose that a single control variable affects the latent processes so $z_i(t)$ is a scalar in this case. Here, setting $R = 0.01$ (as compared e.g., to $R = 10$) would yield a relatively small (large) penalty on the administration of $z_i(t)$, thus allowing the controller to recommend administration of $z_i(t)$ relatively liberally (scarcely). The effects of different choices of these design matrices will be demonstrated later in the context of the illustrative examples.

The basic concept of RHC can be summarized as follows. At the current time t, the optimal control is computed, over a finite fixed horizon from the current time t, to $t+h$. For each t, out of all the control values computed over the time window from t to $t+h$, only the optimal control input values at time t (i.e., $\tau = 0$), denoted as $z_i^*(t)$, is retained and administered to control the state values of the system at time $t+1$. The procedure is then repeated sequentially for the next time, namely, for time $t+1$, over the horizon $[t+1, t+1+h]$, to yield a declining (receding) possible horizon over which future control values may be administered. Since future values of the system are used to determine the optimal current values of $z_i^*(t)$ to be used at a future time point, RHCs are distinct from strict predictions in which past and current values of a system are used to obtain future values of a system's states or associated observed measurements. Our explanations that follow using Pontryagin's minimization principle, are based heavily on (Kwon & Han, 2005, pp. 26–33).

The Pontryagin's minimization principle is broadly used in optimal control theory to find the best possible control inputs for taking a dynamical system's latent states from one time point to another while subjected to particular constraints for the control inputs. This minimization is accomplished by minimizing the Hamiltonian, which summarizes the total discrepancies in the cost function in (8.4) as subjected to some specified constraints. In the absence of process ($\zeta_{SS,i}(t)$) and measurement noises ($\varepsilon_i(t)$), the noise-free version of the state-space model in Equations (8.1)–(8.2) can be used to form one possible Hamiltonian function expressed as

$$\begin{aligned}\mathbf{H} &= (\eta_{SS,i}(t+\tau)-\eta_i^r)'\mathbf{Q}(\eta_{SS,i}(t+\tau)-\eta_i^r)+z_i(t+\tau)'\mathbf{R}z_i(t+\tau)\\ &+p_i'(t+\tau+1)\left(\mathbf{\Phi}_{SS}\eta_i(t+\tau)+\mathbf{G}_{SS}z_i(t+\tau)\right)\text{for } \tau\in 1,\ldots,h-1.\end{aligned} \tag{8.5}$$

This Hamiltonian function can be seen as a Lagrangian function, a type of mathematical function of the form $\mathcal{L}(x,p)=f(x)-pc(x)$, in which $f(x)$ is an objective function and $c(x)$ is a set of constraints. The Lagrangian function helps find the local maximum or minimum of $f(x)$ as subjected to the constraint functions in $c(x)$, and p is the so-called Lagrangian multiplier. In Equation (8.5), $\mathcal{L}(\eta_{SS,i}(t+\tau),z_i(t+\tau))\overset{\Delta}{=}(\eta_{SS,i}(t+\tau)-\eta_i^r)'\mathbf{Q}(\eta_{SS,i}(t+\tau)-\eta_i^r)$ $+z_i(t+\tau)'\mathbf{R}z_i(t+\tau)+c_i(h)$ can be thought of as the per-stage (or per time point) contribution of the cost function of choice at time $t+\tau$. $\mathbf{\Phi}_{SS}\eta_i(t+\tau)+\mathbf{G}_{SS}z_i(t+\tau)$ delineates the constraints imposed by the state-space model of choice on the projected values of $\eta_i(t+\tau+1)$. The vector of Lagrangian multiplier values in $p_i'(t+\tau+1)$ needs to be computed to help solve for the optimal values of $z_i^*(t+\tau)$.

According to the Pontryagin's minimization principle (see pp. 26–33, Kwon & Han, 2005; and Section 3.3.1, Goodwin et al., 2005; Molenaar, 2010), minimization of the cost function in (8.4) requires the following principles to hold, namely:

$$\frac{\partial\mathbf{H}}{\partial z_i(t)}=2\mathbf{R}z_i(t+\tau)+\mathbf{G}_{SS}'p_i(t+\tau+1)=\mathbf{0}, \tag{8.6}$$

$$\frac{\partial\mathbf{H}}{\partial\eta_{SS,i}(t+\tau)}=2\mathbf{Q}(\eta_{SS,i}(t+\tau)-\eta_i^r)+\mathbf{\Phi}_{SS}'p_i(t+\tau+1)=p_i(t+\tau), \tag{8.7}$$

$$\frac{\partial c_i(t+h)}{\partial\eta_i(t+h)}=2\mathbf{Q}_h(\eta_{SS,i}(h)-\eta_i^r)=p_i(t+h), \tag{8.8}$$

The last equation of which is a "boundary condition" that helps determine the value of $p_i(t+h)$ at the end of the control horizon window. Setting Equation (8.6) to **0** and solving for $z_i(t+\tau)$ yields:

$$z_i^*(t+\tau)=-0.5\mathbf{R}^{-1}\mathbf{G}_{SS}'p_i(t+\tau+1).$$

where $z_i^*(t+\tau)$ denotes the optimal control input values that minimize the cost function of interest.

Kwon and Han (2005) derived how the conditions outlined in Equations (8.6)–(8.8) provide the computational equations needed to compute $z_i^*(t+\tau)$ through backward recursions, starting from the end of the control horizon window, $t+h$ to $t+h-1,\ldots,t$ as

$$z_i^*(t+\tau)=-\mathbf{R}^{-1}\mathbf{G}_{SS}'\mathbf{S}_i^{-1}(t+\tau)\left[\mathbf{L}_i(t+\tau+1)\mathbf{\Phi}_{SS}\eta_{SS,i}(t+\tau)+g_i(t+\tau+1)\right] \tag{8.9}$$

where $\mathbf{S}_i(t+\tau) = \left[\mathbf{I}_w + \mathbf{L}_i(t+\tau+1)\mathbf{G}_{SS}\mathbf{R}^{-1}\mathbf{G}'_{SS}\right]$; whereas $\mathbf{L}_i(t+\tau)$ and $g_i(t+\tau)$ are obtained, starting from the end of the control horizon at $\tau = h$ as

$$\mathbf{L}_i(t+h) = \mathbf{Q}(t+h)$$
$$g_i(t+h) = -\mathbf{Q}(t+h)\eta_i^r, \tag{8.10}$$

and then computed recursively backward in time for $\tau = h-1$ to 0 as

$$\mathbf{L}_i(t+\tau) = \mathbf{\Phi}'_{SS}\mathbf{S}_i^{-1}(t+\tau)\mathbf{L}_i(t+\tau+1)\mathbf{\Phi}_{SS} + \mathbf{Q}$$
$$g_i(t+\tau) = \Phi'\mathbf{S}_i^{-1}(t+\tau)g_i(t+\tau+1) - \mathbf{Q}\eta_i^r. \tag{8.11}$$

This controller can be regarded as "deterministic" in the sense that it was originally designed for systems in which perfect knowledge of the latent states is available. In other words, the true values of a system's latent state values are known. In cases where the true state values are subjected to process noises and their true values are unknown, but noisy observed (output) variables are available, $\eta_i(t+\tau)$ may be replaced with predicted or filtered estimates, namely, $E(\eta_i(t+\tau) \mid y_i(1),\ldots,y_i(t+\tau-1))$ or $E\left(\eta_i(t+\tau) \mid y_i(1),\ldots,y_i(t+\tau)\right)$, respectively, from running the Kalman filter algorithm described in Chapter 5 (Molenaar, 2010). Some examples of these variations utilizing alternative latent variable estimator with the LQC were proposed and compared by Chow et al. (2022).

8.2 Illustrative Simulation

To demonstrate the effects of constrained control theory optimization, we simulated bivariate time series for multiple individuals as characterized by the same group-based state-space model. Changes at the latent level are characterized by a vector autoregressive model of lag order 1 with one exogenous variable (i.e., a VAR(1)-X model). At the manifest level, each of these latent processes was identified by one manifest indicator that was subjected further to measurement noise. This model was implemented using the R package, *dynr*, and we provide supplementary code to illustrate how to use functions and by-products from running the Kalman filter using *dynr* to perform control theory optimization.

In state-space form, this VAR(1)-X model with two state variables and two manifest variables can be expressed as

Measurement model:

$$y_{1i}(t) = \eta_{1i}(t) + \varepsilon_{1i}(t)$$
$$y_{2i}(t) = \eta_{2i}(t) + \varepsilon_{2i}(t)$$
$$\left[\begin{array}{cc}\varepsilon_{1i}(t) & \varepsilon_{2i}(t)\end{array}\right]' \sim N\left(\left[\begin{array}{cc}0 & 0\end{array}\right]', \mathrm{diag}\left[\begin{array}{cc}\sigma_{\varepsilon_1}^2 & \sigma_{\varepsilon_2}^2\end{array}\right]\right) \tag{8.12}$$

Dynamic model:

$$\eta_{1i}(t) = \phi_{11}\eta_{1i}(t-1) + \phi_{12}\eta_{2i}(t-1) + g_1 z_i(t-1) + \zeta_{1i}(t)$$
$$\eta_{2i}(t) = \phi_{21}\eta_{1i}(t-1) + \phi_{22}\eta_{2i}(t-1) + g_2 z_i(t-1) + \zeta_{2i}(t)$$
$$\begin{bmatrix} \zeta_{1i}(t) & \zeta_{2i}(t) \end{bmatrix}' \sim N\left(\begin{bmatrix} 0 & 0 \end{bmatrix}', \begin{bmatrix} \psi_{11} & \psi_{12} \\ \psi_{12} & \psi_{22}, \end{bmatrix} \right)$$
$$\begin{bmatrix} \eta_{1i}(1) & \eta_{2i}(1) \end{bmatrix}' \sim N\left(\begin{bmatrix} \mu_{\eta_1 0} & \mu_{\eta_2 0} \end{bmatrix}', \begin{bmatrix} \psi_{110} & \psi_{120} \\ \psi_{120} & \psi_{220}, \end{bmatrix} \right) \tag{8.13}$$

where $\eta_{1i}(t)$–$\eta_{2i}(t)$ are state variables, $y_{1i}(t)$–$y_{2i}(t)$, $z_i(t)$ is the one exogenous or control input used in this illustration, linked to the state variables with corresponding regression coefficients, g_1 and g_2; $\zeta_{1i}(t)$–$\zeta_{2i}(t)$ are process noises for the two state variables, and $\varepsilon_{1i}(t)$–$\varepsilon_{2i}(t)$ are measurement noises.

To implement control theory optimization of the values of $z_i(t)$ for each time point, the values of the parameters, ϕ_{11}–ϕ_{22}, g_1–g_2, $\sigma^2_{\varepsilon 1}$–$\sigma^2_{\varepsilon 2}$, and ψ_{11}–ψ_{22}, all assumed to be known a priori. In the absence of such information, parameter estimation can take place in real time, albeit with greater uncertainty, or they can be estimated a priori using a test data set. In what follows, we demonstrate the latter scenario.

We first generated a test data set using the VAR(1)-X model T = 50 time points, N = 300 individuals, and the following dynamic and measurement parameter values:

$$\phi_{11} = 0.7, \phi_{12} = -0.3, \phi_{21} = -0.2, \phi_{22} = 0.6, g_1 = -0.7, g_2 = -0.4,$$
$$\psi_{11} = 2.0, \psi_{12} = 0.5, \psi_{22} = 1.5, \sigma^2_{\varepsilon_1} = 0.5, \sigma^2_{\varepsilon_2} = 0.5.$$

These parameters were estimated by optimizing the prediction error decomposition function, a log-likelihood function that can be computed using by-products from running the Kalman filter, as described in Chapter 5 and available in the *dynr* package. For data generation, the following initial condition-related parameter values were used:

$$\mu_{\eta 10} = \mu_{\eta 20} = 0, \psi_{110} = \psi_{220} = 1, \psi_{120} = 0;$$

and these specific parameters were assumed known and fixed at their true values. Code for data generation and model fitting to the test data using *dynr* is provided as part of the supplementary code for this chapter.

After estimation of the unknown parameter values using the test data set, we generated another data set consisting of T = 30 time points and N = 3 individuals, using the same true parameter values, initial conditions, and a single exogenous variable, z_{it}, with the value zero for all individuals and all time points. Note that the effects of the control input optimization can be illustrated even in cases involving $N = 1$. Our use of multiple-subject time series simply reflects our intention to demonstrate that once parameter estimates are obtained for the group-based model with reasonably large N and T combination (e.g., with $N \geq 1$ and $T \geq 50$; or $N \geq 100$ and $T \geq 14$), optimal control for multiple individuals can be computed separately for each individual to personalize the timing and dosages of such control theory-driven interventions.

Two randomly selected "shock points" were incurred on each individual's time series of $\eta_{1i}(t)$ and $\eta_{2i}(t)$ within the time window $1 \leq t \leq 25$, with shock magnitudes that were randomly sampled from a uniform distribution in the interval between 10 and 20. This demonstration was motivated by the hypothetical scenario where external perturbations or uncertainties helped propelled individuals to show unusually high values on the state variables (e.g., stress and negative emotions), followed by gradual recovery from the shocks over time to return to the prototypical, stable VAR(1) trajectories that hover around 0, characterized by noise-like fluctuation patterns.

The control input in this case may be a stress reduction strategy of choice, such as exercise or a vacation, that can help lower the stress level as well as negative emotion of an individual. As opposed to devising a "one-size-fits-all" dosage of stress reduction scheme to each individual, we can use the control theory algorithm shown in Equations (8.9)–(8.11) to determine the optimal value of $z_i^*(t)$ that would help accelerate individuals' return to their original VAR trajectories, realized under the scenario with no perturbations.

We applied the following constrained control scheme: we set the control horizon, h, to be 4, and

$$\mathbf{Q} = \mathbf{Q}_h = \mathbf{\Lambda}'_{SS}\mathbf{\Lambda}_{SS} = \begin{bmatrix} 1 & 0 \\ 0 & 1 \end{bmatrix}\begin{bmatrix} 1 & 0 \\ 0 & 1 \end{bmatrix} = \begin{bmatrix} 1 & 0 \\ 0 & 1 \end{bmatrix}; \quad R = 0.001, \text{ or } 1.0.$$

To illustrate the role of selecting different values of R, the cost of administering the control input, we fixed $\mathbf{Q}$ and $\mathbf{Q}_h$ to an identity matrix and varied the value of R from 0.00001 to 10.0. The first value of 0.00001 was relatively small compared to the cost matrices for deviations of the latent variable values from the target trajectory, $\boldsymbol{\eta}_i^r$, selected to emulate the scenario where the costs associated with administering the stress reduction strategy were low, as dictated by having R assume a value that was much lower than those in $\mathbf{Q}$ and $\mathbf{Q}_h$. The value of $R = 1.0$ was larger and identical to the values of $\mathbf{Q}$ and $\mathbf{Q}_h$. In addition, we set $\boldsymbol{\eta}_i^r$ to be 0 (the stable baselines of the VAR(1) process in the absence of external perturbations) for all individuals and time points. We further fixed all the parameters to those estimated using the test data set and *dynr*.

The resultant latent state trajectories, $\eta_{1i}(t)$, for the three hypothetical subjects obtained under the computed optimal control input, $z_i^*(t)$, and the original control input with values of $z_i(t) = 0$ and the group-based state-space model are plotted in Figure 8.1 with R fixed at 0.001 and 1.0. It can be seen

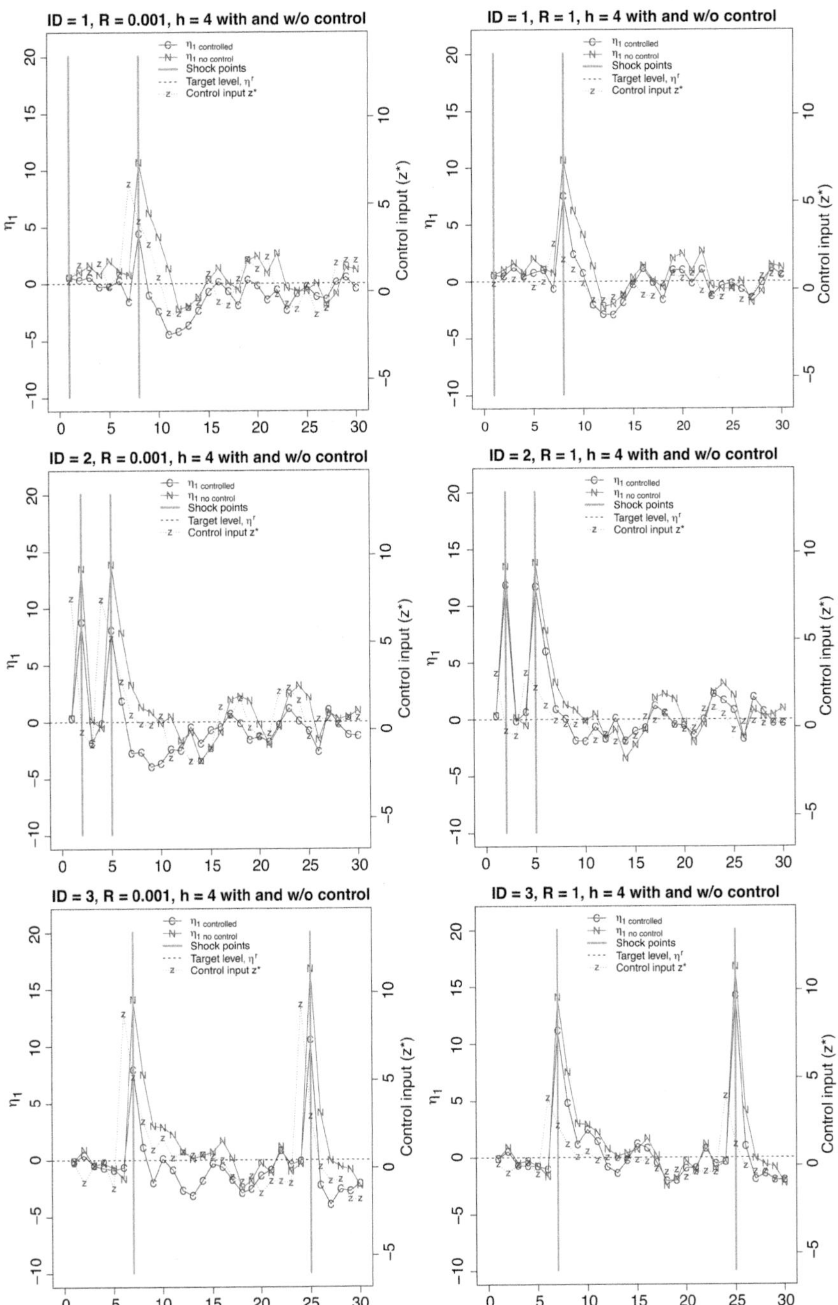

FIGURE 8.1
Simulated trajectories of $\eta_{1i}(t)$ and $\eta_{2i}(t)$ generated with and without the control theory optimized $z_i^*(t)$ with three illustrative subjects, with $R = 0.001$ (left panel) or $R = 1.0$ (right panel), and h, the control horizon window, fixed at 4.

that when the controller was used, higher dosages of $z_i^*(t)$ were automatically diagnosed and delivered around the shock points (the gray, vertical solid lines), and the resultant, "controlled" latent trajectories (see the solid line marked with the symbol "C") clearly show quicker return to the unperturbed trajectories compared with the original trajectories without the control input (see the solid line marked with the symbol "N"). The use of the control horizon window of future time points (i.e., from $t+1,\ldots,t+h$) allows the effects of the shocks to be detected and "remedial" control input values deployed instantaneously at and following the shock points. The values of $z_i^*(t)$ recommended were proportional to the extent of deviations from the target latent trajectories, as expected based on Equation (8.9). For instance, for ID 1, highest dosages of $z_i^*(t)$ were recommended following the second than the first shock point.

In addition, when more "aggressive" control scheme were used (i.e., with lower cost values for R, such as 0.001 as compared to 1.0), highest dosages of control input were recommended, and the system was observed to return to the target level (fixed at 0 across all persons and time points) more efficaciously, after which the dosages of $z_i^*(t)$ were slowly reduced toward 0. To give a concrete example, compare the the plots in Figure 8.2 for ID 1 between $t = 0$ and 10. Notice that under $R = 0.001$ (left panel), as compared to $R = 1.0$ (right panel), higher dosages of $z_i^*(t)$ (the maximum value was slightly lower than 10.0) were deployed during the same time window for the former than the later (the maximum value of $z*_i(t)$ was less than 5.0). Correspondingly, the controlled state trajectory, marked with the symbol "C", also showed less deviations compared to the target level of 0 (the maximum deviation in state value in this time window was approximately 5.0 under $R = 0.001$ as opposed to greater than 5.0 under $R = 1.0$. In fact, the control scheme under $R = 0.001$ actually gave rise to a more pronounced period of "over-corrections" (greater negative deviations in state values) between $t = 10$ and 15) under this scheme compared to the more conservative scheme of $R = 1.0$.

Overall, for all subjects, higher values of $z_i^*(t)$ were suggested around the shock points. When a lower penalty weight was imposed on control input administration (R= 0.001, see the left column of plots, as compared to $R = 1$ in the right column), larger dosages of $z_i^*(t)$ were recommended (e.g., for ID 2, around $t = 2$ and 5; for ID 3, around $t = 5$ and 25), which generally resulted in controlled trajectories with peaks that were closer to the target level of 0. For all subjects, both of the $R > 0$ values considered were able to bring to system closer to its target trajectories than when no control input was administered.

For design purposes, it is of interest to derive some indices for quantifying the costs and benefits associated with using a set of control inputs. Chow et al. (2022) proposed one possible set of cost and benefit functions:

$$\text{Cost\_State} = \sum_{i=1}^{n}\sum_{t=1}^{T-1}\left[(\eta_{SS,i,t} - \eta_{i,t}^{r})'\mathbf{Q}(\eta_{SS,i,t} - \eta_{i,t}^{r})\right] + (\eta_{SS,i,T} - \eta^{r})'\mathbf{Q}_h(\eta_{SS,i,T} - \eta^{r})$$

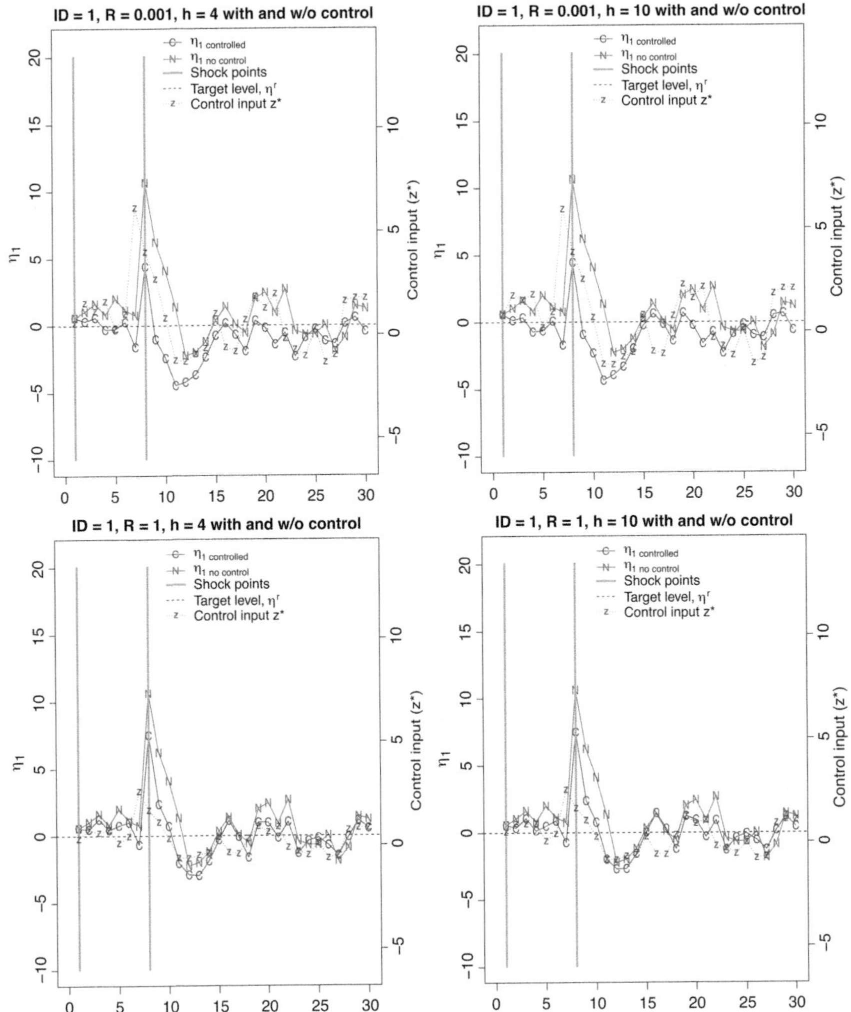

FIGURE 8.2
Simulated trajectories of $\eta_{1i}(t)$ generated with and without the control theory optimized $z_i^*(t)$, with different values of control horizon ($h = 4$ or 10), and the same cost for administering the control input ($R = 0.001$).

$$\text{Cost\_Input} = \sum_{i=1}^{n}\sum_{t=1}^{T} \mathbf{z}'_{i,t}\mathbf{R}\mathbf{z}_{i,t},$$

$$\text{Relative Benefit} = \frac{\text{BaseCost\_State} - \text{Cost\_State}}{\text{BaseCost\_State}},$$

$$\text{Relative Cost} = \frac{\text{Cost\_Input} - \text{BaseCost\_Input}}{\text{BaseCost\_Input}}, \tag{8.14}$$

where Relative Benefit quantifies the change in the quadratic cost associated with state deviations under the current control input scheme relative to the quadratic state cost in a baseline condition, BaseCost_State, (e.g., when no control input is used). Positive (negative) values represent a reduction (increase) in state deviations from the target trajectory compared to baseline. In a similar vein, Relative Cost quantifies the change (specifically, increase) in the quadratic cost associated with the current control input scheme relative to the quadratic input cost in a baseline condition, BaseCost_Input, (e.g., when input values are not determined by the LQC). Positive (negative) values represent an increase (decrease) in input costs compared to baseline.

Using the indices in Equation (8.14), we computed the relative costs and benefits across five different R values, namely, $R = 0.00001$, 0.001, 0.1, 1.0, and 10.0. We generated simulated data for the same set of hypothetical subjects with the same sequences of process noises and shocks, and summarized the root mean squared deviations (RMSDs) of the corresponding state values from their target values (0) as averaged across state variables, all participants, and time points. These values are summarized in Table 8.1.

Notice that with the control input values, the system generally showed smaller RMSDs from the target levels compared to when no control input was applied, regardless of the R values used. That is, the relative benefit values summarized in Table 8.1 are positive across all the R values considered. However, the RMSD and relative benefit values are notably lower at $R = 10$ than at other smaller R values, suggesting that slightly more liberal application of the input (obtained by assigning a lower cost value to the input) is necessary in this case to counteract the effects of the shocks.

Finally, the value of the control horizon (whether $h = 4$ or 10) did not have a salient effect on the values of the computed control input (see Figure 8.2). This may reflect the particular setting of this illustration, specifically, the limited number of imposed shock points and small T designed to facilitate presentation in this chapter. These specific settings may not have provided long enough stretches of deviations in state trajectories to clearly differentiate the consequences of using different choices of the control horizon window.

TABLE 8.1

Mean Squared Deviations, Relative Costs, and Benefits Associated with Different R Values

R	Relative Cost	Relative Benefit	RMSD
0.00001	0.028	0.460	2.478
0.001	2.812	0.441	2.477
0.1	217.663	0.446	2.391
1.0	564.368	0.310	2.410
10.0	187.125	0.055	2.866

Note: RMSD = root mean squared deviation of the state values from the target levels.

Overall, the plots show that even though a group-based state-space model was used with a time- and person-invariant target level (at $\eta_i^r = 0$ for all time points and persons), the optimal control input values computed were indeed person- and time-specific, consonant with the idea behind the call to personalize and better provide targeted educational, health, and myriad other behavioral, cognitive, and clinical changes at times when the changes are most needed at the individual level. Our illustration demonstrated the utility and feasibility of using a controller in conjunction with a group-based state-space model to automate the control of dynamic processes. In terms of computational time, computations were fast and efficient through integration of our constrained control algorithm with the state-space estimation functions from the R package, *dynr* (Ou, Hunter, & Chow, 2019). For instance, iterating through the computations of the control input at three different R values and two values of h for 10 subjects after parameter estimation has been done only took less than one minute on a Mac computer with 2.3-GHz Intel Core i9 and 15 GB of 2400-MHz DDR4 memory.

The illustrated algorithm can be further improved in a number of ways. For instance, the algorithm computes the control inputs in an off-line manner after data from all T are already available. That is, it computes the input values without regard to the real-time effects of applying the control input values on the state variables. Computing the control input values offline can greatly reduce computational time, but the computed input values might show signs of delayed, over- or under-correlations. Alternatives of implementing the computations in real time, possibly in moving windows, are also possible (Chow et al., 2022; Rawlings, 2013).

Even though a stochastic state-space model was used in this chapter as the basis for computing the control input values, the illustrated controller is still an example of a deterministic LQC. That is, the state values were assumed "known and fixed" at the values of Kalman filter estimates. This kind of control scheme works well when the separable principle holds, namely, when optimal control and state estimation can be decoupled under regularity conditions (Alspach, 1975). This was the case in the model considered in the present article, but this assumption may not hold in other empirical scenarios. In such cases, other stochastic control schemes may have to be utilized instead (Alspach, 1975; Bar-Shalom & Tse, 1976; Lu & Zhang, 2016). In addition, Chow et al. (2022) suggested some possible indices for quantifying the relative costs and benefits of administering the control input, and highlighted the importance of evaluating these and other alternative indices in deciding the values of the penalty design matrices, $\mathbf{R}$, $\mathbf{Q}$, and $\mathbf{Q}_h$. In the future, other more formal selection criteria or measures should be considered and further evaluated to help guide the selection of these cost matrices. Finally, the quadratic cost function used in LQCs penalizes positive deviations from the target functions just as heavily as negative deviations. Other alternatives based on nonlinear cost functions may be more relevant in other applications, such as those that aim to penalize deviations more heavily in one direction than the other (e.g., Taguchi, Chowdhury, & Wu, 2005; van den Berg, 2014; Zhang et al., 2014).

8.3 Summary

Technological advances and the widespread use of mobile electronic devices (e.g., smartphones) have provided individuals and practitioners with much richer possibilities for implementing just-in-time interventions, namely, interventions that are tailored to individuals' momentary states and context to maximize gains. The appeal of personalizing the dosages and timing of intervention is clear to many practitioners; however, the burden to make and deliver these personalized decisions can also be costly. In this chapter, we presented and evaluated one LQC, the LQC-State, which automates the computation of optimal input dosage much in the way that the cruise control unit of a car regulates discrepancies between actual and target driving speed. An illustrative example and supplementary code are provided in the context of a bivariate autoregressive model with measurement errors. Even though the algorithm illustrated is just one special variation of the LQCs, we hope that the supplementary code provides a springboard to catalyze more work on personalizing control interventions in future applications.

References

Alspach, D. (1975). A stochastic control algorithm for systems with control dependent plant and measurement noise. *Computers & Electrical Engineering, 2*(4), 297–306. doi: 10.1016/0045-7906(75)90017-8.

Åström, K.J. & Murray, R.M. (2008). *Feedback systems: An introduction for scientists and engineers*. Princeton, NJ: Princeton University Press.

Bar-Shalom, Y., & Tse, E. (1976). Concepts and methods in stochastic control. In C. Leondes (Ed.), (Vol. 12, pp. 99–172). New York, NY: Academic Press.

Bellman, R. (1964). Control theory. *Scientific American, 211*(3), 186–200.

Carver, C.S. & Scheier, M.F. (1982). Control theory: A useful conceptual framework for personality-social, clinical, and health psychology. *Psychological Bulletin, 92*(1), 111–135.

Chow, S.-M., Lee, J., Hofman, A., van der Maas, H.L.J., Pearl, D.K., & Molenaar, P.C.M. (2022). Control theory forecasts of optimal training dosage to facilitate children's arithmetic learning in a digital educational application. *Psychometrika, 87*(2), 559–592.

Goodwin, G., Seron, M.M., & de Don, J.A. (2005). *Constrained control and estimation: An optimisation approach* (1st ed.). London: Springer-Verlag London Ltd.

Kwon, W.H. & Han, S.H. (2005). *Receding horizon control: Model predictive control for state Models*. London: Springer.

Liu, J., Wang, W., Golnaraghi, F., & Kubica, E. (2010). A novel fuzzy framework for nonlinear system control. *Fuzzy Sets and Systems, 161*, 186–200.

Lu, Q. & Zhang, X. (2016). A mini-course on stochastic control. In *Control and inverse problems for partial differential equations* (pp. 171–254). https://www.worldscientific.com/doi/abs/10.1142/9789813276154_0004.

Molenaar, P.C. (2010). Note on optimization of individual psychotherapeutic processes. *Journal of Mathematical Psychology, 54*(1), 208–213. doi: 10.1016/j.jmp.2009.04.003.

Ou, L., Hunter, M.D., & Chow, S.-M. (2019). What's for dynr: A package for linear and nonlinear dynamic modeling in R. *The R Journal, 11*(1), 91–111. doi: 10.32614/RJ-2019-012.

Rawlings, J.B. (2013). Moving horizon estimation. In J. Baillieul & T. Samad (Eds.), *Encyclopedia of systems and control* (pp. 1–7). London: Springer London. doi:10.1007/978-1-4471-5102-9_4-1.

Rivera, D.E., Pew, M.D., & Collins, L.M. (2007). Using engineering control principles to inform the design of adaptive interventions a conceptual introduction. *Drug and Alcohol Dependence, 88*(Suppl 2), S31–S40.

Savi, A.O., van der Maas, H.L.J., & Maris, G.K.J. (2015). Navigating massive open online courses. *Science, 347*(6225), 958–958. doi: 10.1126/science.347.6225.958.

Taguchi, G., Chowdhury, S., & Wu, Y. (2005). *Tagushi's quality engineering handbook.* Hoboken, NJ: Wiley & Sons.

van den Berg, J. (2014). Iterated LQR smoothing for locally-optimal feedback control of systems with non-linear dynamics and non-quadratic cost. In 2014 American Control Conference (pp. 1912–1918). doi: 10.1109/ACC.2014.6859404.

Wang, Q., Molenaar, P., Harsh, S., Freeman, K., Xie, J., Gold, C., & Ulbrecht, J. (2014a). Personalized state-space modeling of glucose dynamics for type 1 diabetes using continuously monitored glucose, insulin dose, and meal intake: An extended Kalman filter approach. *Journal of Diabetes Science and Technology, 8*(2), 331–345. doi: 10.1177/1932296814524080.

Wang, Q., Molenaar, P., Harsh, S., Freeman, K., Xie, J., Gold, C., & Ulbrecht, J. (2014b) Personalized state-space modeling of glucose dynamics for type 1 diabetes using continuously monitored glucose, insulin dose, and meal intake: An extended Kalman filter approach. *Journal of Diabetes Science and Technolology, 8*(2), 331–345. http://www.biomedsearch.com/nih/Personalized-State-space-Modeling-Glucose/24876585.html.

Zhang, J., Li, W., Wang, K., & Jin, R. (2014). Process adjustment with an asymmetric quality loss function. *Journal of Manufacturing Systems 33*(1), 159–165. doi: 10.1016/j.jmsy.2013.10.001.

9

The Intersection of Network Science and Intensive Longitudinal Analysis

9.1 Terminology

A *network* is any *graph* or figure where there are relations among a set of objects. The objects are typically called *vertices* or *nodes*. In the data examples used throughout the book, these would be the variables. The relations among the nodes (or variables) go by many names across the varied disciplines that use network science approaches, with the most general term being *edge*. *Links*, *adjacencies*, *connections*, *ties*, and *paths* are other terms that have been used to describe these relations among nodes. As an example, in Chapter 6, our output from uSEM and VAR analyses were depicted as networks. Here, the nodes were variables, and the edges were the linear relations among variables across time. Figure 9.1 also depicts relations among variables as a network structure. Two nodes that are connected to each other via an edge are considered *neighbors* of each other. Taken together, the matrices used throughout this book that contain lag-0 or lag-1 estimates can be considered as networks, with the variables being nodes and the elements in the matrix being edges.

It may be helpful to clarify the distinction between *network science* and *graph theory*, particularly since the terms "network" and "graph" are often used interchangeably. Both terms refer to considering the data in terms of nodes (or vertices) and edges (links). The subtle difference is that networks often have connections to real-world systems. As some examples, social networks, airline networks, and local area networks all refer to real, tangible concepts. By contrast, graphs refer to mathematical principles characterizing networks. As described in Bondy and Murty (2008), "A graph is an ordered pair (V(G), E(G)) consisting of a set V(G) of vertices and a set E(G), disjoint from V(G), of edges…". Hence the use of these terms tends to be dictated by the subject matter: whether the goal is to advance understanding of real-world phenomena, or theoretical mathematics. Often, graph theoretic metrics are applied to real-world networks, in which case either term is appropriate.

A graph is a visual depiction of a network. Each graph has a corresponding square matrix (network) that contains the values for each edge which could

DOI: 10.1201/9780429172649-9

A. Adjacency Matrix

	irritable	restless	worried	guilty	anhedonia	hopeless	down	concentrate
irritable	1	0.56	0.51	0.16	0	0.19	0.04	0.36
restless	0.56	1	0.54	0.2	0	0.21	0.09	0.38
worried	0.51	0.54	1	0.14	0	0.28	0.17	0.37
guilty	0.16	0.2	0.14	1	0.3	0.36	0.59	0
anhedonia	0	0	0	0.3	1	0.38	0.54	0
hopeless	0.19	0.21	0.28	0.36	0.38	1	0.65	0
down	0.04	0.09	0.17	0.59	0.54	0.65	1	0
concentrate	0.36	0.38	0.37	0	0	0	0	1

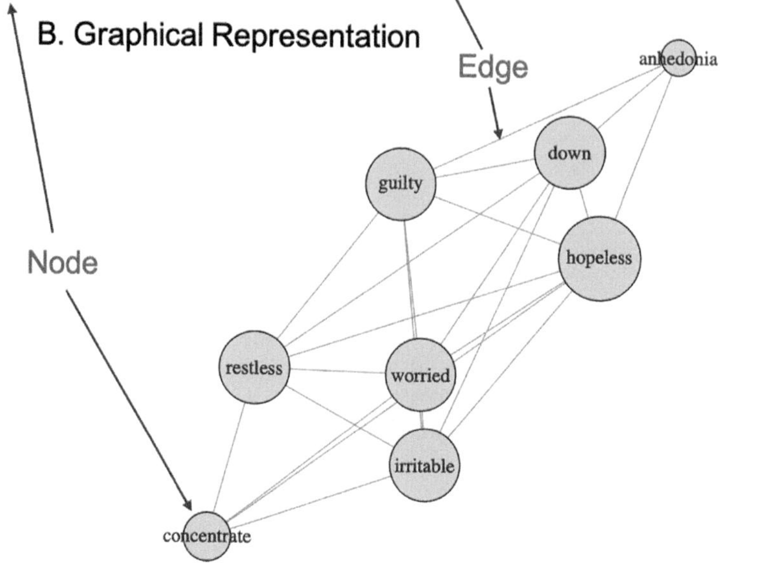

FIGURE 9.1
Depiction of (A) an adjacency matrix, which in this case is a lag-0 correlation matrix for one individual from the Fisher data with negative values thresholded to zero, and (B) the corresponding graphical representation. Note that circles here indicate any type of node (variable), not to be confused with SEM convention to have circles indicate latent variables. Node size corresponds to the degree, or the number of edges in the node.

connect each pair of given nodes. Figure 9.1A depicts this matrix, generally called an *adjacency matrix* in graph theory literature, with the corresponding graph in Figure 9.1B. Specific descriptors are used to define these matrices, some of which match the terminology used thus far in the book. For instance, the adjacency matrices are considered either *directed* or *undirected*, much like we considered dynamic relations as directed or bidirectional/undirected. An example of a directed adjacency matrix would be the $\Phi(u)$ matrix from our VAR models (discussed in Chapters 4–6) which contains the estimates for directed relations at a given lag u. An undirected matrix would be the covariance matrix among residuals in a VAR or hybrid-VAR model (Ψ), or a lag-0 correlation matrix of how variables relate across time. A good rule to remember these by is that directed relations are drawn from edges from asymmetric matrices, whereas undirected relations have edges in symmetric matrices.

Another important descriptor is whether the matrix is *weighted* or *unweighted*. The Φ and Ψ matrices in VAR models are weighted: they provide the estimate for each relation among each pair of nodes, and this estimate is not required to be 0 or 1. An unweighted graph is binary: there is a 0 for an absence of an edge and a 1 for the presence. Our output can be converted into an unweighted graph using thresholding, such as setting values greater than a certain value to 1 and all others to 0, if the aim is to assess the presence or absence of paths. Unweighted graphs are useful to consider when examining data-driven search results as discussed in Chapter 6 to describe the structure of relations. A final descriptor is whether the graph is *sparse* or *dense*. Data-driven or machine learning searches often arrive at sparse graphs, where the values for edges among nodes are dominated primarily by zeros. Dense matrices are those where the edges primarily have non-zero values. Examples of a dense matrix would be correlation, partial correlation, or covariance matrices typically seen in practice where values may near zero but not be exactly zero. Here, each node is connected to each other node, with the edges being the coefficient estimates for each pair of nodes.

Adjacency matrices also go by a variety of terms. A *connection* matrix might represent the geographic connections among train stations or brain regions. In *social networks*, the edges may indicate how many friends a given pair of individuals have in common. Here, the nodes are people. For a *similarity* matrix, the edges represent how similar two given nodes are. This might be useful in the case where nodes are individuals and one seeks to arrive at measures to describe them or cluster the individuals. A *co-occurrence* matrix indicates how often two given behaviors (nodes) co-occurred. This might be used in examining which online behaviors tend to co-occur during a given login session.

As can be seen, there are many different types of matrices and subsequently, many different types of networks. Returning to its use with time series data, one example of how network science is used is when examining functional MRI data. Here, one often obtains correlation (or partial correlation) matrices at a lag of zero and depicts them in network form. Network measures are obtained from these (weighted, undirected, dense) correlation matrices. Another example would be the output obtained from the family of VAR models. GIMME and Graphical VAR output (see Chapter 6) is presented graphically, with nodes being variables and edges being estimates for relations found to exist between a given pair of nodes. One can arrive at network measures from this output to see, for instance, if one variable tends to have a high number of relations to other variables. Yet another example would be to use network approaches to arrive at subsets of variables that tend to covary together across time. As discussed below, the interpretation is similar to cluster analysis, and can be an alternative or complementary piece to the P-technique (see Chapter 3) or other options for factor analysis.

We now turn to network analytic options that are relevant to the output obtained by the time series analysis methods discussed thus far. Throughout, we use the GIMME results from the Fisher data as an example, selecting eight variables that all appear to be negative valence or indicative of a depressed mood: Irritable, Restless, Worried, Guilty, Anhedonia, Hopeless, Down, and Concentrate. Figure 9.2 depicts these results. We begin with network theory measures that summarize the patterns of relations among nodes. This can be used on output obtained from VAR, uSEM (SVAR), hybrid-uSEM, or any component of these such as the sparse lag-0 partial-correlation matrix among residuals obtained in graphical VAR, or the **A** matrix containing estimates for directed lag-0 paths in the uSEM results. In this chapter, we use the **A** and **Φ** matrices from GIMME output. We then turn to community detection, which is a popular technique across many disciplines for clustering nodes into subsets that (in most cases) have a higher value (or weight, depending on values of the network) of within-subset edges relative to the between-subset edges. We show how community detection can be used to subset either variables or people using results

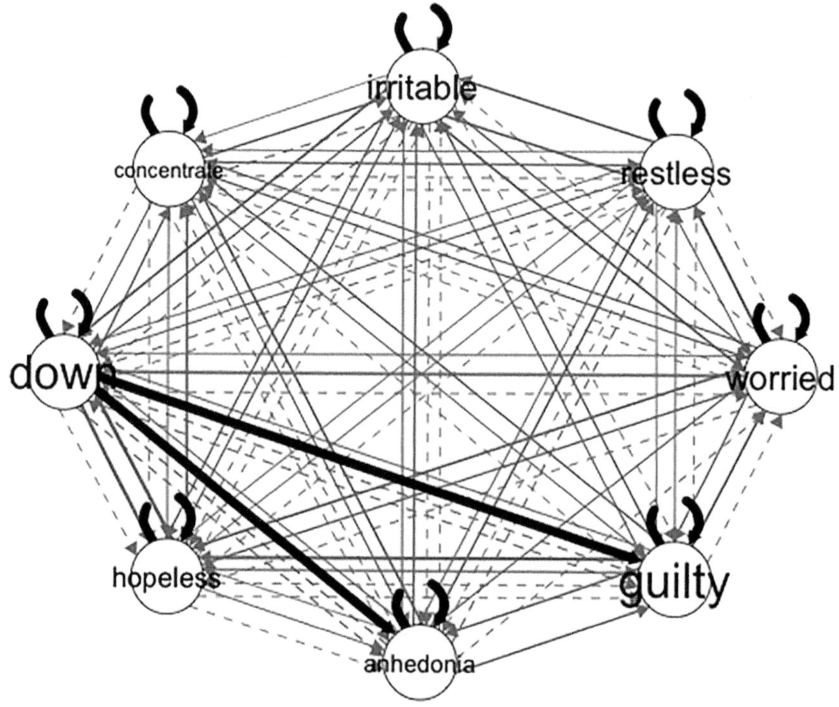

FIGURE 9.2
GIMME output from the Fisher data. Variables are nodes, relations are depicted as directed edges. Dashed lines indicate lag-1 relations; solid lines depict lag-0 relations.

from intensive longitudinal analyses. We close with directions for future work to better enable the use of these methods for the types of data likely obtained by social scientists.

9.2 Network Measures

A number of measures exist for quantifying properties of the overall network and specific nodes. Each one reduces the entire network to one number (or a set of numbers done separately for each node), and in this way can be highly helpful when attempting to make inferences. An example of a global or network-wide attribute would be how integrated the nodes are. Some nodal attributes assess how central a given node is to the overall network, and which nodes tend to have a high number of edges with other nodes. Some measures can be applied both at the nodal and the global levels.

The measures to follow generally consider only positive matrices. This reflects the domains from which the work came. For instance, in social networks, one cannot typically have a negative value reflecting the presence or absence of communication with another person. When considering your own data while reading the below definitions, think critically about what negative values mean for your research question. Sometimes, negative edges are simply a reflection of coding. For instance, there might be a negative edge between a positive valence item and one that is negative. If the negative one were reverse-coded, it would likely be a positive edge. In this case, it might make sense to take the absolute values of edges. If the research question dictates keeping only positive edges, then do so. This can occur in any type of data set. An example is seen in fMRI work, where negative edges are often coded as zero given the interpretation of negative edges here and the preference for quantifying positive edges.

Below, we begin by describing how properties related to edges are quantified. These measures are sometimes used on their own, and also aid in understanding the next measures to be discussed: measures of centrality, segregation, and integration.

9.2.1 Summarizing Edge Values: Degree, Density, Weight, and Strength

Oftentimes researchers wish to obtain one value that can be used to describe the average edge properties of either the network, a subset of nodes within a network, or the nodal level. When looking at overall network edge properties, the summary measure can indicate the overall connectedness, on average, among the nodes. The same type of measures can be used within a subset of nodes to assess how connected they are relative to another subset. Nodal attributes regarding edge properties can indicate how central the node is to the network. The *degree* is the most intuitive and universal measure: it is simply a count of edges. In directed networks, one can look at the *in-degree*, or the count of edges that go into a node (i.e., the arrow ends at a node). *Out-degree* is the opposite: this is the count of edges that go out (i.e., arrow begins at a node) to another node. *Nodal degree* is the number of edges (in or out, undirected or directed) that are associated with that node. *Graph degree* is the total number of edges in a graph. For weighted networks, one can obtain *weight* and *strength*. A specific edge weight, w_{ij}, is the value for a given edge a_{ij} between nodes i and j in the adjacency matrix. Strength (s_i) typically refers to the sum of these edge weights for a given node, with each counted only once:

$$s_i = \sum_{j \neq i} w_{ij} \tag{9.1}$$

We arrived at nodal attributes using the GIMME results on the Fisher data introduced above. Given that some negative values, such as those between Down and Concentrate, might be meaningful, we opted to use the absolute values of the coefficient estimates rather than set negative values to zero. We then added the **Φ** and **A** matrices to arrive at one square matrix per person that quantifies the lagged and contemporaneous edges across variables. Panels A and B in Figure 9.3 depict the degree and strength distributions across the eight variables. One can easily see that degree seems to be highest for the variable Down. This is the overall degree that takes into account both in- and out-degree. The strength for this variable is also consistently higher across individuals. Taken together, one might conclude that down has the greatest number of edges across individuals and also has the largest summed

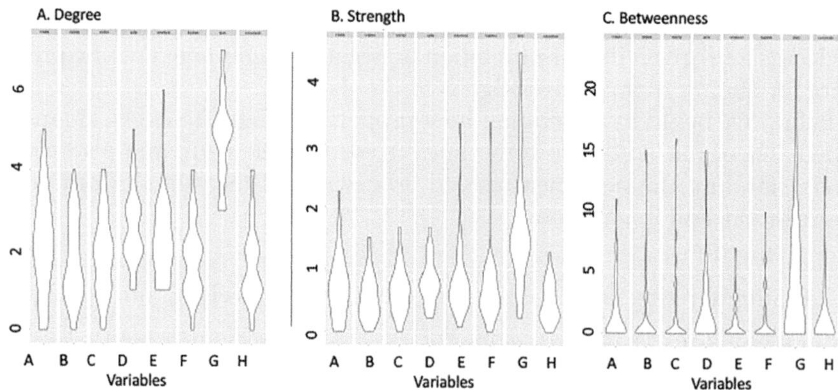

FIGURE 9.3
Distributions for nodal measures. Panel A provides degree or count of edges; Panel B, the strength defined here as the sum of edge weights; and Panel C, betweenness which is the sum of all shortest paths that passes through this node. Variable A = Irritable, B = Restless, C = Worried, D = Guilty, E = Anhedonia, F = Hopeless, G = Down, H = Concentrate.

value of edge strengths. This matches what we see in the GIMME output depicted graphically in Figure 9.2, where two group-level edges originate at the variable Down, with additional individual-level edges evident from and to this variable. The overall *graph weight* is the sum of all unique elements. One can take the weight of the entire network, or subsets of nodes within a network. When we discuss community detection below, we will see how the degree and weight of subsets of nodes can be useful concepts when trying to find subsets of nodes that belong in the same community. For community detection, the goal is often to arrive at a high within-community degree or weight relative to the degree or weight of between-community edges.

Density measures aim to obtain a value that represents how strongly the nodes are interconnected. Again, this can be used for a subset of nodes or the entire network. Density measures average across the nodal edges in some way. Sometimes, this is the average degree across all nodes (Rubinov and Sporns, 2010). A nice example is seen in the work of Bringmann and colleagues (2016), where they averaged across the absolute values of edge weights (rather than a degree) to obtain a measure of network density. First, they arrived at individual-level estimates of sparse VAR models conducted on data collected numerous times over days. The results were put in matrix form, where the nodes were negative and positive emotion variables and the edges were the VAR estimates. Each individual had a matrix (network), and from here three density measures were obtained for each individual: one for all of the negative and positive emotions, one for the subset of negative emotions, and one for the subset of positive emotions. The authors showed that denser networks were associated with higher neuroticism scores, and this was a particularly strong relation when looking just at the density of negative emotions.

9.2.2 Centrality Measures

Both degree and strength have been used at the nodal level to quantify aspects of centrality, or how connected a given node is. Additional measures have been developed that might capture centrality more accurately. Two such measures, *closeness* and *betweenness*, both utilize quantification of the shortest path between each pair of nodes. The shortest path, also referred to as the geodesic distance, is simply the shortest number of edges needed to cross to get from a given node i to another node j. For instance, in Figure 9.1B, we can see that the shortest path between the nodes anhedonia and irritable is 2: anhedonia—hopeless (or down or guilty)—irritable. If two nodes are disconnected, meaning that there is no edge path to follow from a given node to another, then the value of the shortest path for this dyad is infinity.

Closeness centrality for a given node i is the inverse of the average shortest path length from node i to all other nodes. Higher values of closeness centrality indicate more efficient and greater connectivity with other nodes in the network. By taking the inverse, the problem of infinity values is removed as these become zero. *Betweenness centrality* is typically defined as the fraction of all shortest paths in the network that passes through a given node. Sometimes, the sum of the number of times it is the shortest path is used. Intuitively, a node that is part of the shortest path for connecting many other nodes is likely integral to the system.

Figure 9.3C provides the distribution of betweenness for the variables in the Fisher data. It follows from the measures of degree and strength that the variable down also has relatively high betweenness. Importantly, we see that for some individuals, the betweenness of other variables is quite high. These measures taken together help to understand the role the variables play in the system. They can help to assess which nodes help to support the integration of all the nodes in the system, a topic we now turn to.

9.2.3 Measures of Segregation and Integration

Summary information about the edges provides key information regarding general properties. However, these measures do not quantify how connected the nodes are. This may be of interest to a research curious if all of the symptoms tend to be related to each other or not, or in the case of brain networks, if the brain regions segregate into subsets. Such research questions can be probed using measures of segregation and integration.

The *clustering coefficient* reflects the presence of clustered nodes surrounding a given node. This is achieved by counting up the triangles, or the number of neighbors a node has that are also neighbors with each other. Defining tri_i as the number of triangles around node i, we have

$$tri_i = \frac{1}{2}\sum_{j,h} a_{ij}a_{ih}a_{jh} \tag{9.2}$$

where a values here are zero if the two nodes do not share an edge and 1 if they do. The clustering coefficient is the fraction of a given node's neighbors that are also neighbors of each other (Watts and Strogatz, 1998):

$$c_i = \frac{1}{N}\sum_{i}^{N} \frac{2tri_i}{e_i(e_i - 1)} \tag{9.3}$$

where c_i is the clustering coefficient and degree for node i. e_i represents the number of edges in the neighborhood of i. An average can be taken for all nodes in the network to reflect expected clustered connectivity among nodes in the network.

Integration measures attempt to capture how well-connected all nodes are, even those that are not directly connected to each other (in a sparse network). The average shortest path length, or *characteristic path length*, is one such measure. Higher values indicate less integrated networks. *Global efficiency* (Latora & Marchiori, 2001) is the average inverse shortest path, and may be a better measure since it is primarily influenced by short paths (Achard & Bullmore, 2007; Rubinov & Sporns, 2011).

System segregation is commonly used in brain networks where subsets of specific brain regions are expected to have high correlations with each other and lower correlation coefficient estimates with those outside the subset. It can be used for any set of nodes where researchers wish to see if predefined subsets have higher average edge weights with others in their cluster versus nodes outside the cluster. Specifically, system segregation is the average within-cluster edge strength minus the between-cluster edge strengths, divided by the average within-cluster edge strength. These clusters of nodes can be found in a data-driven manner, which we now turn to.

9.3 Community Detection Algorithms

Community detection is an umbrella term for a large number of methods that partition nodes into subsets, clusters, or subgroups of nodes, referred to as *communities*. Typically, the aim is for the nodes within a community to have a higher degree or weight (depending on the properties of the network and data) with each other than with nodes outside their community (Fortunato, 2010). Conceptually, the output from community detection is similar to what is obtained in cluster analysis, which also has a long history in the behavioral sciences and may be more familiar to the reader. In both cases, a vector of community (or cluster) assignments is obtained, and nodes with the same labels are considered to be closer to each other. Both approaches provide unsupervised classification. As with cluster analysis, there are many methods for producing the clusters and criteria to optimize in the general category of community detection approaches. The primary difference between cluster analysis and community detection is the data being used—historically cluster analysis has referred to the clustering of variables, whereas community detection has roots in the clustering of people from a network perspective. The first acknowledged application of community detection was the use of similarities in voting patterns to cluster individuals in small political bodies (Rice, 1927). More recently, community detection has been used on networks with various qualities, from correlation matrices representing connections among brain regions to social networks of individuals. The community detection methods all have one thing in common: they require a square matrix (i.e., network). As we will show later in the chapter, community detection in the intensive longitudinal context can be used to subset individuals who have similar dynamic models or to subset variables at the individual level that relate to each other, providing an alternative to P-technique.

The development of numerous community detection techniques was sparked by a seminal paper written by physicists Girvan and Newman (2002) on a community detection technique geared toward both social and biological sciences. The majority of these algorithms come from physics, computer science, nonlinear dynamics, sociology, and discrete mathematics (Fortunato, 2010), with emerging methods coming out of psychology (e.g., Hoffman et al., 2018). The large number of available algorithms makes a review that is both concise and exhaustive impossible. However, some generalizations can be made. First, most methods were developed with big data in mind and were not always tested on small data sets. Second, most were developed initially for use with unweighted graphs, although some also adapted a weighted version (although most do not allow for negative values). Third, most were developed for sparse graphs.

Fourth, many use a measure called "modularity" to arrive at the optimal solution. Modularity is a score that indicates the degree to which edge values with others within a community are high relative to edge values with nodes between communities. It is a relative measure, meaning that it can be used to compare solutions with the same data but cannot be used to compare solutions from different networks. As such, it cannot be used as an absolute measure—there is no cutoff for what a "good" solution is. Since its inception, it has been known that random graphs can be high in modularity (Porter, Onnela, & Mucha, 2009), again providing caution when using modularity scores as an absolute measure of segregation. As a relative measure, it is extremely helpful in arriving at the optimal solution for community detection algorithms by comparing solutions for the same network.

In weighted networks, modularity measures the strength of edges within a community compared to the strength of edges outside the community (Blondel et al., 2008). When the strength of edges within a group is larger than what is expected at random given the properties of the nodes, modularity rises (Newman, 2004; Newman & Girvan, 2003). Formally, modularity for weighted networks can be written as (Fan et al., 2007)

$$Q^w = \frac{1}{2w} \sum_{ij} \left(w_{ij} - \frac{w_i w_j}{2w} \right) \delta(s_i, s_j) \tag{9.4}$$

with δ being 1 if the nodes belong to the same community and 0 if otherwise. The value $w_i w_j / 2w$ provides the expected strength of the edge between the given nodes with w being the summation of the edge strengths in the network and w_i (w_j) the summed strength of node i (j). w_{ij} is the edge strength for the provided node pair.

For the purposes of this book, we focus on one community detection approach, Walktrap (Pons & Latapy, 2006), which is currently the most popular method used in psychology. Walktrap is versatile in its use and is one of few methods that perform well on graphs of varied characteristics, ranging from unweighted to weighted, large to small, and sparse to dense (Gates et al., 2016; Golino & Epskamp, 2017).

9.3.1 Walktrap

Walktrap has been used on various types of networks capturing human behaviors, psychology, and biology. For instance, it has been used to identify subsets of psychological symptoms that tend to covary together across people (e.g., Martarelli et al., 2020) as well as subsets of individuals who have similar dynamic processes (e.g., Wright et al., 2019). It also has been used extensively in functioning neuroimaging data to arrive at subsets of brain regions that covary across time (e.g., Ferreira & Zhao, 2016).

As with many other community detection algorithms, Walktrap first recasts the network into a matrix that measures the similarities between nodes. Walktrap does so by generating a matrix of transition probabilities. Each element in this matrix reflects the probability of going from one node to each given node in a random walk of a given length. One starts with the transition probability of going from node i to node j at each step: $p_{ij} = {}^{a_{ij}}/_{d_i}$, where d_i is the degree for node i and a_{ij} is the (i,j)th element of the adjacency matrix, $\mathbf{A}$. Powers of the transition matrix $\mathbf{P}$ drive the process, with the probability of going from node i to node j at a random walk of t being p_{ij}^t. The algorithm rests on the property that the ratio of the probabilities of going to from i to j and from j to i through a random walk of length t depends only on the degrees of these nodes:

$$d_i p_{ij}^t = d_j p_{ji}^t. \tag{9.5}$$

This information guides attempts at arriving at communities. First, two nodes in the same community will have a high transition probability. This is because nodes that are all highly connected to each other will have a high probability of going to each other in a relatively short number of walks. Secondly, p_{ij}^t is influenced by the degree d_j; thus, high degree nodes tend to have higher transition probabilities. It must be noted that a high transition probability does not guarantee they are in the same community. Nodes that have a high degree or centrality may have high transition probabilities (relative to other nodes) with numerous nodes, and will be placed with nodes for which it is the highest. Finally, nodes in the same community relate to other nodes in a similar way. That is, they tend to have shared neighbors.

Another important property concerns the length of t. As the length of the random walk goes toward infinity, the probability of being on a given node j depends only on the degree of j and not the starting node. For this reason, t cannot be too long as information regarding the structure of the network would be lost.

A traditional clustering algorithm, Ward's method (Ward, 1963), is then conducted on this transition matrix. Ward's algorithm is agglomerative, meaning it starts with each node in its own community and iteratively combines nodes based on which combination would minimize some sum of squares.

Here, the distances between two given nodes are minimized. First, we define the distance between two nodes at a given random walk length t:

$$r_{ij} = \sqrt{\sum_{k=1}^{N} \frac{\left(p_{ik}^{t} - p_{jk}^{t}\right)^{2}}{d_{n}}} = \left\| \mathbf{D}^{-1/2} P_{i\bullet} - \mathbf{D}^{-1/2} P_{j\bullet} \right\| \quad (9.6)$$

where $\|\cdot\|$ is the Euclidean norm and $\mathbf{D}$ is a diagonal matrix of the degrees with $\mathbf{P} = \mathbf{D}^{-1}\mathbf{A}$. This provides the L^2 distance between the two probability distributions. This is the first distance minimized to create communities. Next, each community is iteratively combined according to the minimization of the distance between them. Defining the probability of going from community C to node j in t steps as $p_{Cj}^{t} = \frac{1}{|C|} \sum_{i \in C} p_{ij}^{t}$, the distance between two communities can be found as:

$$r_{C_1C_2} = \sqrt{\sum_{k=1}^{N} \frac{(p_{C_1k}^{t} - p_{C_2k}^{t})^{2}}{d_{k}}} = \left\| \mathbf{D}^{-1/2} P_{C_1\bullet}^{t} - \mathbf{D}^{-1/2} P_{C_2\bullet}^{t} \right\| \quad (9.7)$$

With these distances in hand, one can choose which two communities to merge at each step k based on the minimization of the mean squared distances between each node and its community:

$$\sigma_{k} = 1/N \sum_{C \in \mathcal{P}_k} \sum_{i \in C} r_{iC}^{2} \quad (9.8)$$

which is the squared distance between each node and its community. For greater computational efficiency, a variation of σ is used that would be induced if C_1 and C_2 were merged into $C_3 = C_1 \cup C_2$:

$$\Delta\sigma(C_1, C_2) = \frac{1}{2}\left(\sum_{i \in C_3} r_{iC_3}^{2} - \sum_{i \in C_1} r_{iC_1}^{2} - \sum_{i \in C_2} r_{iC_2}^{2} \right) \quad (9.9)$$

where paired communities with the lowest valued of $\Delta\sigma$ would be merged. Starting from the minimization of distances for all nodes into the first communities, the algorithm continues to combine communities in this manner resulting in a dendrogram of solutions. The solution with the highest modularity Q is often selected (although different criteria can be used).

We will use Walktrap in the analyses to follow. First, we demonstrate how Walktrap can be used to subset individuals based on their patterns and estimates of dynamic models. Here, the nodes in the network are individuals. Next, we compare the use of Walktrap to P-technique and demonstrate how community detection can supplement or complement this traditional measurement approach.

9.4 Using Community Detection to Subgroup Individuals with Similar Dynamic Processes

In the time series context, we can use community detection to subset individuals who have similar patterns and estimates of dynamic relations. This has a number of benefits: first and foremost, it may help aid in making inferences. Interpreting the heterogeneity in results across individuals can be difficult, particularly as the sample size N increases. Say a large minority of individuals have one edge between two nodes x and y, and you also see that a large minority of individuals have an edge between a different set of nodes x and z. One *post hoc* question might be, are the people who have a relation between the first two nodes the same people who have a relation between the second two nodes? The interpretation of results might differ greatly based on this information. If they are the same people, then we could make statements regarding potential indirect effects or parent status (i.e., if a given node x explains variance in both nodes y and z). However, if node x relates to node y for some individuals and node z for a subset of other individuals, then perhaps it is not a hub or particularly important node for any subset of individuals.

Individuals can be subsetted into subgroups (or communities, or clusters) based on the patterns and estimates in their dynamic networks (e.g., matrices of SVAR estimates). Generating subgroups, or clustering individuals, serves two general goals that may coincide: (1) arrive at distinct subsets of individuals, who differ substantially in their patterns (i.e., are different) and this is validated with an outside measure or (2) arrive at subsets of individuals with some differences in their patterns that are consistently present within the subgroup, but perhaps they still share a number of similarities with individuals in other subgroups. The best practice is to determine this *a priori* on the basis of explicit hypotheses, and thus have a clear goal for the analysis. For instance, Dajani and colleagues (2019) recently sought to identify if categorical models of neural substrates were valid. That is, they sought to identify if there were distinct subgroups that were different enough in terms of their patterns to represent entirely different types of brain processes. Their findings suggest that, while individuals differ in their patterns, they cannot be clustered into robust subgroups based on the relations found for the brain regions chosen.

In other work, researchers align their hypotheses in terms of the second goal of clustering. That is, they make no strong statements regarding how different individuals must be, but rather are curious to identify which relations (if any) tend to co-occur for subsets of individuals. For instance, Wright and colleages (2018) utilized subgrouping to better detect signal from noise in order to get more precise recovery of paths. One of the main goals was to see the frequency with which specific paths were found, and patterns of relations that might tend to co-exist in order to better understand common symptom manifestation. Here, it is not necessary that the subgroups be robustly different from each other in terms of their patterns of relations. Rather, the subgrouping of individuals is used to help understand and better capture the varied individual-level processes.

9.4.1 Exemplar Method: Subgrouping GIMME

Community detection approaches offer an ideal set of algorithms for the unsupervised classification of individuals. One method that subgroups individuals based on patterns and estimates of dynamic models is Subgrouping GIMME (S-GIMME; Gates et al., 2017). Here, rather than using the final individual-level models, subgrouping occurs within the GIMME algorithm (see Chapter 6) following the group-level search. From here, one must construct a network. The network used here will have individuals as the nodes and some sort of measure of similarity in dynamic models as the edges. Community detection of any sort requires decisions regarding features. In the context of subgrouping based on individuals' dynamic models, features here refer to how we wish to quantify the similarity between individuals in terms of their patterns (and perhaps also estimates) of relations.

To quantify similarity the S-GIMME algorithm counts the number of relations that each pair of individuals has in common defined as being significant after Bonferoni correction as well as having the same sign (i.e., positive or negative). The estimates included in the count come from two sources: one being the individual-level estimates of the group-level paths following the group search, and the other source being the Expected Parameter Change (EPC) estimates. EPCs indicate which paths, after controlling for group-level path estimates, might be selected for each individual as the algorithm moves to the individual-level search. Each dynamic path (i.e., a lagged or contemporaneous relation) that is both significant and has the same sign for two individuals is included in the count of similar paths. The total count for each pair becomes the edge for that element in the similarity matrix (see Figure 9.4B). The resulting similarity matrix used for subsetting individuals is a symmetric count matrix where the nodes are people and the edges similarity counts. By default, some sparsity is induced by subtracting the lowest value from all elements in the matrix. Another option includes fixing a specific proportion of elements to zero, or conducting a grid search to identify the proportion of zeros that leads to the lowest modularity (see *gimme* package documentation; Lane et al., 2022).

The approach described here can be adapted for any output from a model in the DFM family of approaches. Here, we focus on S-GIMME, given its ease in use and public availability. The use of count matrices in S-GIMME was informed by simulation studies showing this method for assessing similarity provided more reliable subgroup results than other methods, such as taking the correlation between two individuals' path estimates. Hence future work working with different types of dynamic models may also want to consider the use of similarity count matrices.

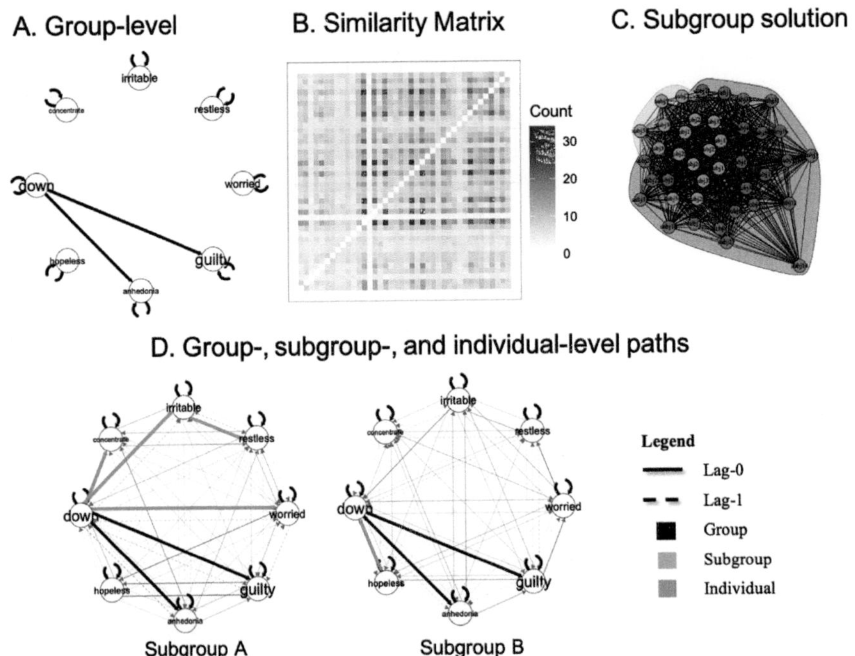

FIGURE 9.4
S-GIMME results on Fisher data. (A) Results after a group-level search; (B) similarity matrix; (C) depiction of Walktrap results; (D) graphs of final results for two subgroups. Solid lines represent contemporaneous (lag-0) relations, and dashed lines represent lag-1 relations. Line width corresponds to the proportion of individuals with that path.

Next one must choose which community detect algorithm to use. S- GIMME provides the option to use any method available in the *igraph* package (Csardi & Nepusz, 2006). Since Walktrap provides reliable recovery of community assignments when the input matrices are either correlation or count matrices (Gates 2016; Golino & Epskamp, 2017), and the evaluation of S-GIMME investigated similarity matrices of both forms, Walktrap was the algorithm evaluated in the Monte Carlo simulations of Gates and colleagues (2017). It performed exceptionally well, providing excellent recovery of subgroups when the count similarity matrix was used.

The complete algorithm is as follows. The group-level search for paths that are present across the majority of individuals is conducted as in regular GIMME. Then, the subgroup stage begins as depicted in Figure 9.4A. Next, the similarity matrix is obtained. Individuals are then placed into subgroups by conducting Walktrap on the sparse count matrix. From here, subgroup-level paths are obtained by conducting the same algorithm used to arrive at group-level paths. However, the search for paths is only conducted among those within the same subgroup. For all subgroups, the group-level paths are used as the null or baseline model from which the search proceeds.

That is, paths are added at the subgroup level if they are detected in the majority of individuals within the subgroup after accounting for group-level paths.[1] Once the subgroup-level paths are obtained for each subgroup, the individual-level path search commences. The group- and subgroup-level paths are the starting point for the individual-level searches. Lane and colleagues (2017) found that including subgroup-level paths, which detect signal from noise by using shared information much like the group-level paths, improves upon the reliable recovery of the individual-level paths. In the end, the algorithm provides three levels of paths: group, subgroup, and individual levels.

9.4.2 Community Detection Empirical Example: Identifying Subsets of Individuals

We conducted S-GIMME on the Fisher variables introduced previously in the chapter. Readers can replicate this analysis using the online code (available through https://www.routledge.com/9781482230598). Figure 9.4D contains the graphical representations of the relations found. Note that the group-level relations between Down with Guilty and Anhedonia still exist (as also seen in Figure 9.2), which is expected given that the subgrouping stage occurs after the group-level stage. Two subgroups were found, one with N_1=24 and the other N_2=16.

Some interesting results emerged. It appears that those who have a relation between Positive and Anhedonia (Subgroup 1) tend to also have a relation between Positive and Irritable. Since the direction goes from Positive to these two other attributes, it seems that having information about one's self-reported level of Positive may explain variance in Irritable and Anhedonia after controlling for other effects. This could be an area for clinical exploration, as perhaps Positive is a hub symptom that, if increased, may alter the feelings of Irritable and Anhedonia (although causality would have to be examined experimentally). We also see that for Subgroup 1 many individuals have an indirect effect between Positive and Avoid People that seems to be mediated by levels of Anhedonia. For all individuals, the relation between Positive and Anhedonia was negative in this subgroup, and the relation between Anhedonia and Avoid People was positive. Finally, this subgroup of individuals evidenced a covariance between Positive and Energetic. This suggests that both attributes have a cause that is outside of the system of variables included here.

Subgroup 2 evidenced some differences, most notably the lack of subgroup-level relations found for the first Subgroup. It is important to note that some of the paths seen in Subgroup 1 exist for Subgroup 2 at the individual level. This is not an uncommon finding, and simply reflects that some individuals may have a few things in common with the other subgroup, but overall were

[1] A different subgroup cutoff threshold can be used for what constitutes the majority, and 51% has often been used since 75% may become too strict due to rounding if the number of individuals within a subgroup is small.

more similar to the subgroup in which they were placed. For this group, the only path that existed for the majority of the subgroup was from Energetic to Positive. Exploration of the estimates for this effect across individuals reveals that the relation was positive for all individuals in this subgroup. This suggests that at times when the individuals reported higher levels of Energetic, they are also likely to have higher levels of Positive after controlling for other variables in the individuals' models.

9.5 Assessing Robustness of Community Detection Solutions

Community detection solutions do not have fit indices that can be used to assess the quality of the solution or how well the nodes separate into their respective communities. Even solutions from a community detection algorithm that has been shown to provide reliable results may not provide robust communities if the data simply do not have high segregation. However, there are techniques that can be used to evaluate if the solutions are robust or not. One way to assess this is to see if small perturbations to the network result in different subgroup classifications for individuals. A second common way is to see if the modularity value obtained on the original network is higher than expected by chance when compared to random networks (i.e., networks with no community structure) that have similar properties. Here we outline methods for assessing the robustness of sparse count matrices.

9.5.1 Obtaining Random Networks

Before explaining these methods we need to define a random network as this is used in both methods. The Erdos–Renyi (ER) binary random matrix in particular is often used as a point of comparison to other graphs. Edges in ER matrices are equiprobable across all nodes. In this way, it provides a null contrast for comparison against matrices that might have nodes that segregate. A weighted extension of this provides sets of edges randomly drawn from the distribution of node strengths (defined previously). Weighted random matrices must have edge weights drawn from the same distribution with equal probability across nodes. Garleschi (2009) introduced a method for arriving at random weighted count matrices that is weight-preserving. That is, the overall weight of the original matrix is maintained while randomness is Induced. In the following, what is often seen in network literature across numerous types of data, the distribution of edges is assumed to be negative binomial with parameters $N - 1$ and $1 - p$, where N is the number of individuals and p a constant. The probability of a given edge weight occurring between any two given nodes is:

$$e(w) = p^{w}(1-p). \tag{9.10}$$

Using the definition of weight introduced above, where the network weight is defined as the sum of the weights of all unique edges, we can find the optimal criterion for tuning p:

$$p^* = \frac{2W}{N(N-1)+2W}. \tag{9.11}$$

One can randomly arrive at a new value for a given edge weight from this distribution using Bernoulli trials with success probability p^*. For an edge randomly selected to be perturbed, the edge value will first be set to zero. Then, Bernoulli trials commence for that edge using the success probability of p^*. If the first trial is not successful, then the edge weight remains "0". If the first trial is successful, the edge is set to "1". For each subsequent successful Bernoulli trial, a "1" is added to this edge weight value until an unsuccessful trial occurs. As edges are independent in count matrices, we can alter the edges without needing to worry about downstream impact.[2]

The rewiring of the network can be done incrementally to induce increasing levels of randomness in the original adjacency matrix. We can gradually increase the degree of perturbation, or α, beginning with 1% of the edges all the way to 100% of edges. When $\alpha = 1.00$ and all of the edges have been perturbed, we obtain a completely rewired network with a random distribution of edges. It is unlikely that this random network would have a community structure as each node is equally likely to have the same edge weight with every other node.

9.5.2 Approach 1: Identifying When Solution Changes

We can assess the robustness of community detection solutions obtained from count matrices by comparing solutions from these increasingly randomized graphs with the solution from the original. The steps are as follows:

1. Randomly perturb a given number of edges according to α.
2. Conduct community detection on this partially rewired network.
3. Compare the solutions (i.e., the nodal assignments or labels) of the perturbed network.

These steps are repeated numerous times across a given α, and then across a range of α to identify when the solutions start to differ. If the solutions when 20% ($\alpha = 0.20$) of edges are perturbed are more similar to the original solution than when 20% of the nodal assignments are randomly changed, then the solution is said to be robust. If the solution changes greatly when α is low, meaning that nodal assignments differ even with relatively few edge changes, then the solution likely is not robust. This analysis is carried out using the package *perturbR* (Gates et al., 2019).

[2] This is not true of correlation matrices, where altering one element will influence all other elements due to the triangular property.

Referring to Figure 9.5 will aid in understanding the process. On the x-axes, we have the proportion of random rewiring or perturbation, and on the y-axes, we have two measures of similarities in community detection solutions: the variation of information (VI; Meilă, 2007) and the adjusted rand index (ARI; Hubert & Arabie, 1985). The VI has previously been used to compare cluster assignments obtained from different data sets such as those described here (Karrer et al., 2018). Drawing from concepts of information and entropy, the VI considers how much information each cluster solution has and how much information one cluster solution provides about the other.

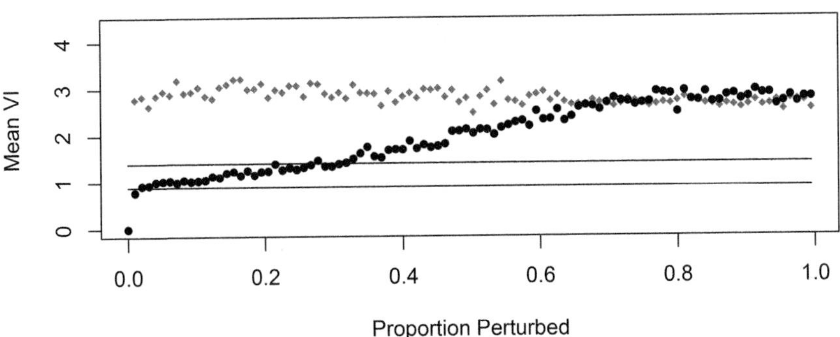

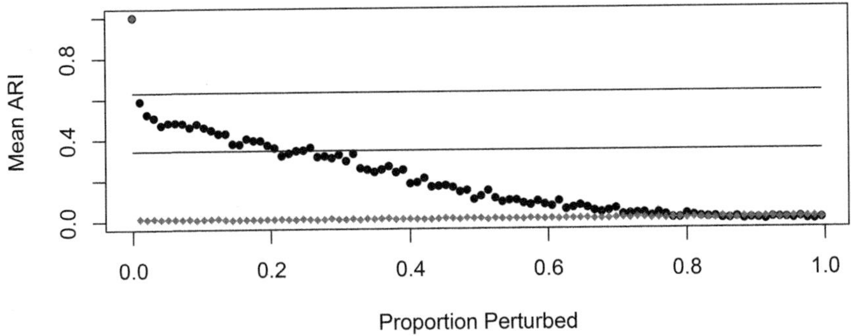

FIGURE 9.5
perturbR robustness results on Fisher data. Diamonds indicate mean values from random networks; circles indicate the mean values from networks with specific proportion perturbed (α). (A) Variation of information (VI) results with higher values indicated a greater difference in community solutions. The top and bottom lines indicate VI when 20% and 10% of nodes are placed in randomly different communities, respectively. (B) Adjusted rand index (ARI) results with lower values indicated a greater difference in community solutions. The top and bottom lines indicate ARI when 10% and 20% of nodes are placed in randomly different communities, respectively.

Given a total of K clusters in a given partition $\mathcal{P}$ with community assignments in vector C, the probability of being in any one cluster k is:

$$P(k) = \frac{n_k}{N} \tag{9.12}$$

where n_k provides the total number of nodes in cluster k and N the total number of nodes in the matrix. The uncertainty or entropy of the vector of cluster assignment probabilities is

$$H(C) = \sum_{k=1}^{K} P(k) log P(k) \tag{9.13}$$

The entropy value depends on the relative proportions of the cluster assignments, with zero uncertainty occurring only when there is one cluster. The mutual information of the cluster solutions, or how much information one solution contains about the other solution, can be provided as follows:

$$P(k,k') = \frac{|C_k \cap C'_{k'}|}{N} \tag{9.14}$$

The prime indicates results from the second community solution. Of note, the cluster in the second solution need not have the same label as the first or the same number of clusters. Hence, $P(k,k')$ provides the probability of overlap of assignments for any given pair of cluster labels found in the two solutions. With this information the mutual information associated with the two clusterings C,C' can be computed:

$$ICC(C,C') = \sum_{k=1}^{K} \sum_{k'=1}^{K'} P(k,k') log \frac{P(k,k')}{P(k)P(k')} \tag{9.15}$$

Taking together the individual entropies associated with the two cluster solutions $H(C)$ and $H(C')$ and the mutual information that each provides for the other ($I(C,C')$) one can arrive at the VI:

$$VI(C,C') = HC(C) + HC(C') - 2I(C,C') \tag{9.16}$$

The VI is not bound by any limits, and higher values indicate greater differences in the community detection solutions. While being a relative measure may limit the interpretability by requiring the solutions be compared in a relative rather than absolute sense, the VI has properties that make it superior. Namely, the VI has been proven to be a true metric by obeying the triangular inequality. This means that it has all the properties of a proper distance measure. Since there is no absolute cut-off for when two solutions differ according to the VI, we must compare solutions to something.

First, we visually compare it to the VI values obtained when community solutions from completely random graphs are compared with the original solution (red diamonds). Second, we can identify when the VI value is above what we would expect the VI to be when 20% of the nodes are randomly placed into different communities (i.e., given different membership labels). This is the top line in Figure 9.5A.

We see evidence of robustness in the solution. The value for the top line here is 1.40 (although given the random nature, this may shift on different runs). The average VI for the iteratively perturbed matrices reaches this mark when $\alpha = 0.28$, indicating that 28% of edges need to be randomized before the solution becomes as different from the original as when 20% of the nodes are randomly put into communities. This suggests a solution that is robust to minor perturbations.

The Hubert-Arabie adjusted rand index (ARI_{HA}; Hubert & Arabie, 1985) provides an alternative measure. It assesses the similarity of node cluster assignments while taking into account what would be expected by chance. It measures the proportions of nodes consistently clustered together in two given community assignment solutions regardless of the arbitrary labels in the solutions.

The ARI_{HA} is computed as follows:

$$ARI_{HA} = \frac{\binom{N}{2}(a+d) - [(a+b)(a+c)+(c+d)(b+d)]}{\binom{N}{2}^2 - [(a+b)(a+c)+(c+d)(b+d)]} \tag{9.17}$$

where each pair of nodes provides a count for either a, b, c, or d. The value a indicates the number of pairs placed in the same community for both the true and recovered partition. Both b and c indicate the wrong placement of nodes, with the former indicating the counts of pairs in the same group for the true community structure but different groups for the recovered structure (and the latter indicating the opposite). Finally, d indicates the count of pairs that are in different communities in the generated data and also different communities in the recovered structure. The ARI_{HA} has an upper limit of 1.0, which indicates perfect recovery of the true community structures, and lower values indicate incrementally poorer recovery. The R package *clues* can be used to calculate ARI_{HA} (Chang et al., 2015).

We can see that the bottom line in Figure 9.5B, which is the ARI_{HA} when 20% of the nodes are randomly placed into different communities, hits the average ARI_{HA} for perturbed matrices when $\alpha = 0.20$. This is right on the cusp of our criterion. Taken together, these two results suggest the community detection solution is borderline robust. One benefit of the ARI_{HA} relative to the VI is that it has an upper limit of 1.00, which indicates that the two solutions are identical. The ARI_{HA} can go below 0.00, in which case it is usually set to zero. Hence it allows for an absolute rather than relative indication of similarity in community solutions. A drawback is that it can be downward biased (Karrer et al., 2008; Meilă, 2007).

9.5.3 Approach 2: Evaluating Modularity

Another way to see if the solutions are robust is to see if the modularity value obtained from the original solution is higher than expected from random networks with the same weight. As modularity is a relative and not an absolute index, there is no distribution, for instance, for identifying if the value is significantly different than zero. When this is the case we can create a distribution. For this, we simply take repetitions of weight-preserved random networks where $\alpha = 1.00$. Then, we look at the distribution of modularity from this set of repetitions and identify if the modularity value in the original solution is higher than the 95 percentile of the modularity values obtained in these repetitions.

Using the same similarity matrix as above obtained from S-GIMME, we can now investigate if the modularity value was higher than we'd expect by chance. Surprisingly, given the above results with VI and ARI, it is not—the value obtained in the original matrix of $Q^w = 0.044$ is not higher than 95% of the modularity values obtained on random graphs with the same weight (0.054). Figure 9.6 depicts the distribution and the original modularity value. Thus we have conflicting evidence regarding the robustness of our solution.

9.6 Community Detection and P-Technique

Community detection has a longer history in neuroscience than in the behavioral sciences and is widely used in functional MRI studies. Here, the algorithms have been used to arrive at subsets of brain regions that tend to covary together at a lag of zero. The first introduction of this class of techniques to psychology literature offered it as analogous in some ways to latent variable models (Cramer et al., 2010). Much like P-technique (introduced in Chapter 3) models observed variables that relate to specific latent variables, community detection can arrive at subsets of observed variables that tend to correlate with each other. The two share some commonalities: both often use the covariance or correlation matrix as inputs. That is, pseudo-ML estimation of P-technique attempts to arrive at a good-fitting model implied covariance matrix of observed variables, whereas community detection can similarly be run on the covariance or correlation matrix of observed variables—but here it is referred to as a weighted, undirected network. Both approaches provide information regarding which set of variables tend to covary together, as well as assign labels to the observed variables indicating which factor or community it belongs to.

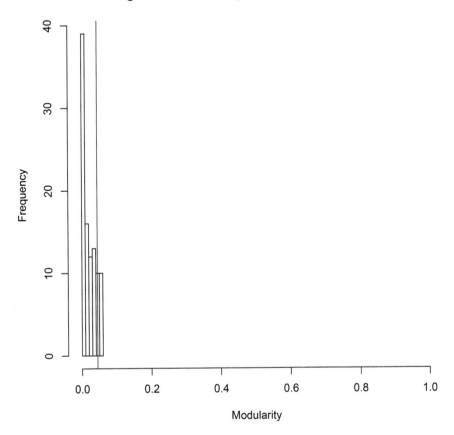

FIGURE 9.6
Distribution of modularity scores on weight-preserved random graphs with similar properties as the similarity matrix obtained from the Fisher data in S-GIMME analysis. The modularity obtained in the original solution (red line) is not above the value seen for the upper 5% of the random samples.

Emerging evidence suggests that, for both small (e.g., 25 variables) and large (e.g., 1000) data sets community detection can recover the factor model structure used to generate the data. (Gates et al., 2016; Golino & Epskamp, 2017). A benefit that community detection may have is that the algorithms perform well on large data sets (i.e., a large number of nodes), whereas P-technique, as estimated using most SEM approaches, is limited in the number of variables that can be used. Thus community detection may aid in exploratory P-technique by helping to identify the number of factors in the data, or used instead of P-technique when the aim is to arrive at subsets of observed nodes that covary without consideration of measurement error. Both studies cited here found that Walktrap performed well in terms of recovering the correct number of factors in the data generating model as well as the nodal assignment (i.e., which node belonged to which factor). Hence, to arrive at the number of factors an additional option to traditional exploratory factor analysis or the method described in Chapter 3 on P-technique would be to use community detection on these variables. After arriving at a model for which nodes relate to the same community, one can return to the dynamic factor analytic framework to incorporate lagged relations and potential cross-lags in the model if they wish, or continue analyses in the observed-variable space.

While both P-technique and community detection can be used to gain information regarding which subsets of nodes tend to covary together, it is important to keep in mind that many differences exist between the two approaches. First, the estimation approaches are vastly different. Specifically, P-technique estimated with quasi-ML seeks to obtain a model-implied covariance matrix that is similar to the observed covariance matrix, whereas community detection optimizes the clustering of nodes into distinct subsets based on some network characteristics of choice, such as a correlation or partial correlation matrix. P-technique can be either exploratory or confirmatory or some combination of both, whereas community detection is entirely exploratory. Factor analyses provide a number of statistics such as fit indices and estimates for how much variability is, in a given variable, is explained by the latent construct, and analogous measures exist in community detection. Community detection only allows one label per variable, whereas factor analyses allow for a variable to load on two or more latent variables (i.e., have more than one label).

The interpretation also differs in some ways. P-technique, as with other latent variable approaches, supposes that the latent construct causes or gives rise to variability in the indicators, with variables often considered to be conditionally independent of this latent construct. The emphasis on P-technique is attempting to obtain a quantitative measure of a hidden construct that is difficult to observe or measure directly. This is in contrast to community detection, where each observed variable is assumed to have no measurement error and not to be caused by the community or latent construct. Rather, the observed variables give rise to each other (in a directed network) or have common causes outside of the system of variables (in the undirected case such as correlation matrices). These differences are important to keep in mind when selecting an analytic plan.

9.6.1 Community Detection Example: Identifying Subsets of Variables

For this example, we use the one individual from the Fisher data: participant 2, who was in subgroup 2 in the above analyses. We first conducted an exploratory P-technique using the *psych* package. The EBIC and TLI point to a two-factor solution (see online supplemental code) with the rotated solution providing a good fit (TLI = 1.00, RMSEA = 0.00). The factor loadings, presented in Table 9.1, suggest that Guilty, Anhedonia, Hopeless, and Down

TABLE 9.1

Factor Loadings for Fisher Data

Variable	Factor 1	Factor 2
Irritable	–0.01	**0.72**
Restless	0.04	**0.77**
Worried	0.13	**0.70**
Guilty	**0.58**	0.10
Anhedonia	**0.56**	–0.19
Hopeless	**0.65**	0.16
Down	**1.00**	–0.06
Concentrate	–0.10	**0.51**

Note: Values greater that 0.50 are in bold.

have the same primary factor, with Irritable, Restless, Worried, and Concentrate having higher loadings on the second factor.

We conducted Walktrap on the lag-0 correlation matrix for this participant. Negative values were thresholded to zero. The variables were separated in a similar manner to what was seen in the exploratory P-Technique approach: guilty, anhedonia, hopeless, and down all belonged to the same community, with the other variables in the second community. Figure 9.7 depicts the solution as well as the corresponding unthresholded correlation matrix. One can see from the matrix the strong relations among within-community nodes, and weaker or negative values in the between-community nodes.

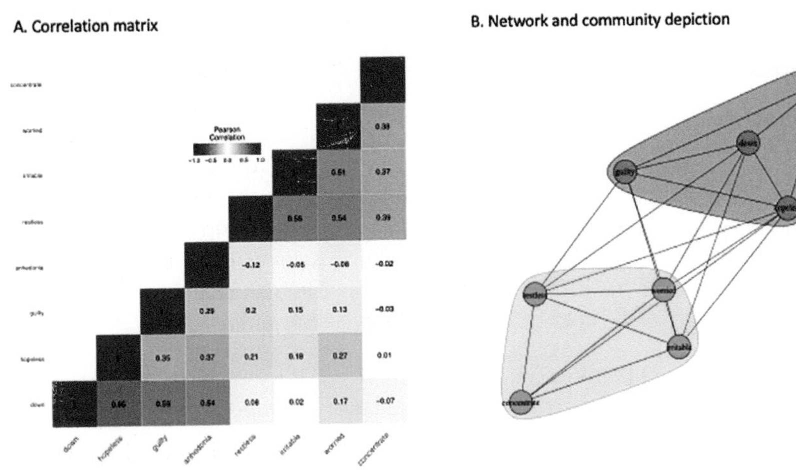

FIGURE 9.7

Depiction of community detection results on data with a two-factor solution. (A) Unthresholded correlation matrix for individual number 2. (B) Corresponding network depiction and community solution. Red lines indicate edges between communities; black lines are within communities. Red and blue colored circles enclose nodes in the same community.

9.7 Discussion

In this chapter, we introduced network science terminology as well as measures that have come from this literature. The measures can be applied to the output from analyses within the DFM framework, ranging from VAR models to lag-0 correlation matrices (of observed values or residuals). These measures provide alternative insights into how variables or individuals are interrelated either at a lag, contemporaneously, or combined.

We also discussed at length some approaches for community detection, an approach for subsetting nodes into subgroups or communities. We showed how one can do this to subgroup individuals using their time series analyses output. We also demonstrated how we could arrive at a subset of nodes that correlate highly with each other, and demonstrate the similarity to P-technique results.

This chapter is intended to serve as a first introduction for the intersection of network science approaches and intraindividual analyses of humans. The field is quickly adopting these methods for use with data having the qualities seen in empirical human studies. For instance, researchers have begun to combine latent variable and network analysis, allowing for the benefits of each approach to be capitalized upon (Gates, Fisher, & Bollen, 2020; Epskamp, Rhemtulla, & Borsboom, 2017). Other innovations will surely arrive soon that take into account such issues such as small matrix sizes, non-normally distributed data, and other features seen in data on human behaviors, self-reports, and psychophysiological data.

References

Achard, S. & Bullmore, E. (2007). Efficiency and cost of economical brain functional networks. *PLoS Computational Biology, 3*(2), e17.

Blondel, V.D., Guillaume, J.-L., Lambiotte, R., & Lefebvre, E. (2008). Fast unfolding of communities in large networks. *Journal of Statistical Mechanics: Theory and Experiment, 2008*(10), P10008. doi: 10.1088/1742-5468/2008/10/P10008.

Bondy, J.A., Murty, U.S.R., et al. (1976). *Graph theory with applications* (Vol. 290). London: Macmillan.

Bringmann, L.F., Pe, M.L., Vissers, N., Ceulemans, E., Borsboom, D., Vanpaemel, W., Tuerlinckx, F., & Kuppens, P. (2016). Assessing temporal emotion dynamics using networks. *Assessment, 23*(4), 425–435.

Chang, F., Carey, V., Qiu, W., Zamar, R.H., Lazaarus, R., & Wang, X. (2015). clues R package version 0.6.2.2 [Computer software]. https://cran.r-project.org/web/packages/clues/index.html.

Cramer, A.O., Waldorp, L.J., Van Der Maas, H.L., & Borsboom, D. (2010). Comorbidity: A network perspective. *Behavioral and Brain Sciences, 33*(2–3), 137–150.

Csardi, G. & Nepusz, T. (2006). The igraph software package for complex network research. *International Journal of Complex Systems, 1695*(5), 1–9.

Dajani, D.R., Burrows, C.A., Nebel, M.B., Mostofsky, S.H., Gates, K.M., & Uddin, L.Q. (2019). Parsing heterogeneity in autism spectrum disorder and attention-deficit/hyperactivity disorder with individual connectome mapping. *Brain Connectivity, 9*(9), 673–691.

Epskamp, S., Rhemtulla, M., & Borsboom, D. (2017). Generalized network psychometrics: Combining network and latent variable models. *Psychometrika, 82*(4), 904–927.

Fan, Y., Li, M., Zhang, P., Wu, J., & Di, Z. (2007). Accuracy and precision of methods for community identification in weighted networks. *Physica A: Statistical Mechanics and Its Applications, 377*(1), 363–372.

Ferreira, L.N. & Zhao, L. (2016). Time series clustering via community detection in networks. *Information Sciences, 326*, 227–242.

Fortunato, S. (2010). Community detection in graphs. *Physics Reports, 486*(3–5), 75–174. doi: 10.1016/j.physrep.2009.11.002

Garlaschelli, D. (2009). The weighted random graph model. *New Journal of Physics, 11*(7), 073005.

Gates, K.M., Fisher, Z.F., Arizmendi, C., Henry, T.R., Duffy, K.A., & Mucha, P.J. (2019). Assessing the robustness of cluster solutions obtained from sparse count matrices. *Psychological Methods, 24*(6), 675.

Gates, K.M., Fisher, Z.F., & Bollen, K.A. (2020). Latent variable GIMME using model implied instrumental variables (MIIVs). *Psychological Methods, 25*(2), 227.

Gates, K.M., Henry, T., Steinley, D., & Fair, D.A. (2016). A Monte Carlo evaluation of weighted community detection algorithms. *Frontiers in Neuroinformatics, 10*, 45.

Gates, K.M., Lane, S.T., Varangis, E., Giovanello, K., & Guiskewicz, K. (2017). Unsupervised classification during time-series model building. *Multivariate Behavioral Research, 52*(2), 129–148.

Girvan, M. & Newman, M. (2002). Community structure in social and biological networks. *Proceedings of the National Academy of Sciences, 99*(12), 7821–7826. doi: 10.1073/pnas.122653799

Golino, H.F. & Epskamp, S. (2017). Exploratory graph analysis: A new approach for estimating the number of dimensions in psychological research. *PLoS One, 12*(6), e0174035.

Hoffman, M., Steinley, D., Gates, K.M., Prinstein, M.J., & Brusco, M.J. (2018). Detecting clusters/communities in social networks. *Multivariate Behavioral Research, 53*(1), 57–73.

Hubert, L. & Arabie, P. (1985). Comparing partitions. *Journal of Classification, 2*(1), 193–218. doi: 10.1007/BF01908075.

Karrer, B., Levina, E., & Newman, M.E. (2008). Robustness of community structure in networks. *Physical Review E, 77*(4), 046119.

Lane, S.T. & Gates, K.M. (2017). Automated selection of robust individual-level structural equation models for time series data. *Structural Equation Modeling: A Multidisciplinary Journal, 24*(5), 768–782.

Lane, S.T, Gates, K.M, Fisher, Z., Arizmendi, C., Molenaar, P., Hallquist, M., & Gates, M.K.M. (2022). Package 'gimme: Group Iterative Multiple Model Estimation' [Computer software]. https://cran.r-project.org/web/packages/gimme/index.html.

Latora, V. & Marchiori, M. (2001). Efficient behavior of small-world networks. *Physical Review Letters, 87*(19), 198701.

Martarelli, C.S., Bertrams, A., & Wolff, W. (2020). A personality trait-based network of boredom, spontaneous and deliberate mind-wandering. *Assessment*. doi: 1073191120936336.

Meilä, M. (2007). Comparing clusterings – An information based distance. *Journal of Multivariate Analysis, 98*(5), 873–895.

Newman, M. (2004). Detecting community structure in networks. *The European Physical Journal B, 38*, 321–330.

Newman, M. & Girvan, M. (2003). Finding and evaluating community structure in networks. *Physical Review E, 69(*2), 26113. doi:10.1103/physreve.69.026113

Pons, P. & Latapy, M. (2006). Computing communities in large networks using random walks. *Journal of Graph Algorithms and Applications, 10*(2), 191.

Porter, M.A., Onnela, J.-P., & Mucha, P.J. (2009). Communities in networks. *Notices of the AMS*, 1082–1097. http://arxiv.org/abs/0902.3788

Rice, S.A. (1927). The identification of blocs in small political bodies. *The American Political Science Review, 21*(3), 619–627. doi:10.2307/1945514.

Rubinov, M. & Sporns, O. (2010). Complex network measures of brain connectivity: Uses and interpretations. *NeuroImage, 52*(3), 1059–1069. doi:10.1016/j.neuroimage.2009.10.003.

Ward, J. H. (1963). Hierarchical grouping to optimize an objective function. *Journal of the American Statistical Association, 58* (301), 236–244. doi:10.1080/01621459.1963.10500845.

Watts, D.J. & Strogatz, S.H. (1998). Collective dynamics of 'small-world' networks. *Nature, 393*(6684), 440–442.

Wright, A.G., Gates, K.M., Arizmendi, C., Lane, S.T., Woods, W.C., & Edershile, E.A. (2019). Focusing personality assessment on the person: Modeling general, shared, and person specific processes in personality and psychopathology. *Psychological Assessment, 31*(4), 502.

Index

I

K

L

M

N

O

P

Q

R

S